Ground Engineering Equipment and Methods

Frank Harris

GRANADA
London Toronto Sydney New York

Granada Technical Books
Granada Publishing Ltd
8 Grafton Street, London W1X 3LA

First published in Great Britain by
Granada Publishing

British Library Cataloguing in Publication Data

Harris, Frank
    Ground engineering, equipment and methods.
    1. Civil engineering
    I. Title
    624   TA145

ISBN 0-246-11239-5

Printed and bound in Great Britain by Richard Clay
(The Chaucer Press) Ltd., Bungay, Suffolk

Contents

Preface

During the past ten to fifteen years national economies have been subjected to rapidly increased trade flows, resulting in keener competition in the design, production and distribution of goods. The construction industry has not been immune from these effects, as demonstrated by the need for both designers and contractors to seek work in the international market place. Traditionally, however, the education of civil engineers and builders has largely concentrated on theory and design, leaving the newly qualified to gain knowledge of production methods piecemeal on construction sites. Consequently, many aspects of our technology are no longer in the vanguard of developments. Thus the author has become aware of the demand for practical reference books dealing with modern construction practice.

This third volume in the series on construction plant has been specifically written for students on undergraduate civil engineering and construction courses and for recently qualified engineers and builders. Similarly it will find a market amongst technicians in training at colleges and other institutions. The treatment is directed towards ground engineering and foundation construction. A special effort has been made to present the principles for deciding on methods of construction, devising temporary works and selecting plant and equipment. The text is fully augmented by diagrams and tables, and performance data are included to help the reader prepare plans and cost estimates of work.

Finally, a book of this type would not be possible without the assistance of many individuals and organisations. Much of the technical information was provided by construction equipment manufacturers and these have been listed in the bibliography at the end of each chapter. Also several of my former students in the Department of Civil Engineering at Loughborough University of Technology have provided supplementary material.

I am especially indebted to Professor R. Seeling, Professor G. Pohle, Messrs W. Sinemus, U. Muller and P. Seller at the Aachen Technische

Hochschule, West Germany, who provided the inspiration and much information for the writing of this book during my study leave in 1976. In addition, other material made available at Aachen T. H. by Professor W. Leins and Professor W. Jureka, now at Vienna T. H., was particularly helpful.

Acknowledgements

The author is especially grateful to friends and colleagues for their helpful comments on some of the specialist topics in the book. The book has also benefited from project studies undertaken by former students, particularly N. R. Short, P. A. Hopgood, E. J. Ellerton, K. F. Cook and F. P. Lewis. Thanks are extended to Janet Redman who performed the arduous task of redrafting the original diagrams and sketches and to Mrs E. M. Lincoln for locating much diffused published material. Finally I wish to thank Vera Cole for cheerfully and intelligently assuming the wearisome task of typing the manuscript.

CHAPTER ONE

Introduction

The job of the construction engineer is primarily concerned with devising construction methods, designing and erecting temporary works and managing labour and plant. Consequently there is need for a construction-oriented education to augment undergraduate studies, which usually concentrate on theory and design.

An important aspect of such an education is adequate knowledge of the ground engineering methods fundamental to the successful execution of most civil engineering and building works. All too often expensive delays occur because construction procedures below ground level have not been carefully considered and proper temporary works provided.

The purpose of this book therefore is to set out the information required by a recently qualified engineer or builder in selecting appropriate construction techniques. The temptation in presenting information of this kind, however, is to follow the case study approach. This has been deliberately avoided because construction projects often vary in nature, with each usually requiring a different emphasis in the choice of construction methods. For example, the factors favouring selection of a de-watering system in certain conditions may not apply in another soil type or location, where a coffer-dam may be more appropriate. Thus construction techniques have been grouped and classified under like headings, with particular attention given to diagrammatic illustration, in order that the fundamental procedures may be easily appreciated and alternatives considered.

Each chapter of the book deals with a broad area of construction technology, and within each, specific construction principles and methods are separately described. Expected production output rates and criteria for plant selection are provided to assist in preparing alternative schemes.

The chapters include:

Compressed air supply

Compressed air is used to drive many items of construction plant, including drills, tunnelling equipment, pile hammers, etc. Principles and practices of production and distribution of compressed air supplies are explained and described.

Rock drilling

Percussive, rotary-percussive, 'down-the-hole', rotary and diamond drilling methods and equipment are described and compared. Criteria in selecting drilling rigs, bits and flushing mediums are set out and expected rates of penetration tabulated.

Soft ground drilling

Continuous flight augering, intermittent augering, and grabbing methods are described. Drilling bits for application in soil to soft rock are compared and the need for and installation of protective borehole linings is explained. Performance characteristics for each are enumerated.

Grouting

A theory of grout penetration is developed and methods of designing grouting patterns are illustrated. Appropriate types of grout for use in soils and rocks are recommended. Selecting grouting equipment and methods, estimating grout quantities and *in situ* testing for permeability are explained.

Ground de-watering

Principles and theory of pumping from wells and sumps are explained. Numerical examples illustrate methods of calculating the size of wells and well layouts. Installing procedures for deep wells, well points, vacuum wells, recharge wells and sumps are described.

Explosives and rock blasting

The theory of explosions is analysed and formulae for calculating quantities of explosive for use in quarrying, tunnelling and demolition work are derived. Types, selection and use of explosives are described including fuse and electrically-detonated systems.

Tunnelling methods

Classical, shield and rotary machine methods are examined. Propping systems and the functions and installation of linings are explained. Consideration is given to pipe jacking and thrust boring techniques used for small diameter tunnels and sewers. Rates of advance, dealing with ground water and bad ground, are highlighted. Reference is also made to submerged tubes and compressed air working.

Pile-driving and sheet piling

Theoretical formulae are derived to aid selection of pile hammer sizes. Drop, single-acting, double-acting, diesel, hydraulic and vibratory pile hammers are described and rated. Piling equipment and techniques are compared, including pile caps and helmets, dollies, leaders and frames. Methods of installing and extracting sheet piles are given particular attention, including work over water.

Shoring systems

Open excavations, trench timbering, sheeting, linings, slide rails and drag boxes are described and compared, with special emphasis given to output rates and labour and plant requirements. The construction and application of single- and double-walled sheet piled coffer dams, piled walls, diaphragm walls, injected membranes, and ground anchors are described and illustrated. The design, construction and sinking procedures for caissons, including box, open and pneumatic types are discussed and explained.

Soil classification and identification

Many of the topics in the book require an understanding of soil and rock mechanics and properties.

Therefore, before continuing further it is perhaps prudent to define some of the terms in the text as they may differ from descriptions used in codes of practice. Indeed it seems unlikely that a precise standard terminology will ever be universally recognised.

The civil engineer commonly refers to soil and rock. Soil is a naturally occurring material composed of mineral grains of various types and sizes. A major feature of a soil is the ability to separate the grains by agitation in water. Rock, however, is a much stronger material which

Table 1.1 Soil types, characteristics and terminology

	Soil description	Particle size (mm)	Method of assessing compaction/strength	
	Boulders	200–250	difficult to determine in the field	non-cohesive
	Cobbles	60–200		
gravel	coarse	20–60		
	medium	6–20	Loose – can be excavated with a spade or 50 mm wooden peg can be easily driven	
	fine	2–6		
sand	coarse	0.6–2.0	Compact – requires a pick for excavation and 50 mm wooden peg is hard to drive	
	medium	0.2–0.6		
	fine	0.06–0.2		
silt	coarse	0.02–0.06	Soft – easily remoulded in the fingers	cohesive
	medium	0.006–0.02	Firm – can be remoulded by strong pressure in the fingers	
	fine	0.002–0.006		
clay	very fine only	less than 0.002	Very soft – extrudes between fingers when squeezed in the hand Soft – moulded by light finger pressure Firm – moulded by strong finger pressure Stiff – very difficult to mould Hard – can be indented by thumb nail and is brittle	
	peats and organic clay, silts and sands	varies	Firm – fibres compressed together Spongy – very compressible and open structure Plastic – can be moulded in hand and smears the fingers	

Descriptions applied when secondary constituents are present

Description	Content of secondary soil	Term (secondary–primary)
slightly	up to 5%	e.g. slightly silty sand
—	5% to 15%	e.g. silty sand
very	15% to 35%	e.g. very silty sand
and	35% to 50%	e.g. clay and silt

Table 1.2 Main rock types and characteristics

Rock description		Compressive strength (N/mm^2)	Rock type
very weak weak moderately weak	soft	under 1.25 1.25–5 5–12.5	Shales, mudstones, poorly compacted sedimentary rocks, highly weathered igneous rocks, boulder clay
moderately strong strong	medium	12.5–50 50–100	Sandstone, limestone, most metamorphic weathered igneous rocks
very strong	hard	100–200	Quartzite, granites, most extrusive igneous rocks
extremely strong	very hard	200+	Fine-grained granites, most intrusive igneous rocks and horn felsites

Note: These are generalisations; sometimes a soft rock may be tougher and more difficult to penetrate than a hard rock.

can generally only be broken up by the application of strong crushing forces. Rocks are described as igneous, sedimentary or metamorphic according to how they were formed.

In the United Kingdom a widely accepted classification system for soils and rocks can be found in BS 5930:1981 – *Code of Practice for Site Investigations.* In the U.S.A. there are two systems. These are the Unified Soil Classification System generally used by engineers and government agencies and the AASHO (American Association of State Highway Officials) System commonly adopted for road design and construction. Elements of these systems are summarised in Tables 1.1 and 1.2 with some slight changes in terminology from the official text.

Many readers will have undertaken quite comprehensive courses as students, and have a good understanding of fundamental soil mechanics; for others, further reading may be necessary. Thus a carefully selected bibliography is provided at the end of each chapter. Furthermore, the reading list at the end of this section is designed to provide an introductory guide to soil mechanics and foundation engineering. In particular *Soil Mechanics in Engineering Practice* by Terzaghi and Peck and *Foundation Design and Construction* by Tomlinson are recommended.

Bibliography

Antill, J.M. and Ryan, P.W.S. (1967) *Civil Engineering Construction*. Angus & Robertson, Sydney.

Atkinson, J.H. (1981) *Foundations and Slopes*. McGraw-Hill, London.

Atkinson, J.H. & Bransby, P.L. (1978) *The Mechanics of Soils*. McGraw-Hill, London.

Bazant, Z. (1979) *Methods of Foundation Engineering*. Elsevier, Amsterdam.

Bell, F.G. (1981) *Engineering Properties of Soils and Rocks*. Butterworths, London.

Bolton, M. (1979) *Soil Mechanics*. Macmillan, London.

Craig, R.F. (1974) *Soil Mechanics*. Van Nostrand Reinhold, New York.

Day, P.E. (1973) *Construction Equipment Guide*. Wiley & Sons, New York.

Hanna, T.H. (1973) *Foundation Instrumentation*. Trans. Tech. Publications, London.

Havers, J.A. & Stubbs, F.W. (1971) *Handbook of Heavy Construction Equipment*. McGraw-Hill, New York.

Holmes, R. (1975) *Introduction to Civil Engineering Construction*. College of Estate Management.

Jaeger, J.C. & Cook, N.G.W. (1979) *Fundamentals of Rock Mechanics*.

Chapman & Hall, London.
Jumikis, A.R. (1971) *Foundation Engineering*. Intertext Educational Publishers, Scranton, U.S.A.
Lambe, T.W. (1979) *Soil Mechanics*. Wiley & Sons, New York.
Legget, R.F. (1962) *Geology and Engineering*. McGraw-Hill, New York.
Nunnally, S.W. (1980) *Construction Methods and Management*. Prentice-Hall, N.J.
Peck, R.B. et al. (1973) *Foundation Engineering*. Wiley & Sons, New York.
Peurifoy, R.L. (1970) *Construction Planning Equipment and Methods*. McGraw-Hill.
Scott, C.R. (1980) *An Introduction to Soil Mechanics*. Applied Science Publishers.
Seelye, E.E. (1956) *Foundations*. Wiley & Sons, New York.
Smith, R.C. (1976) *Principles and Practices of Heavy Construction*. Prentice-Hall, N.J.
Szechy, K. & Varga, L. (1978) *Foundation Engineering*. Akademiai Klado, Budapest.
Terzaghi, K. & Peck, R.B. (1967) *Soil Mechanics in Engineering Practice*. Wiley International, New York.
Tomlinson, M.H. (1975) *Foundation Design and Construction*. Pitman, London.
Wilun, Z. (1972) *Soil Mechanics in Foundation Engineering*. Intertext., Scranton, U.S.A.
Winterkorn, H.F. (1975) *Foundation Engineering Handbook*. Van Nostrand Reinhold, New York.
Wood, S. (1977) *Heavy Construction*. Prentice-Hall, N.J.
Wu, T.H. (1976) *Soil Mechanics*. Allyn & Bacon, London.

CHAPTER TWO

Compressed air

Introduction

Compressed air is required on construction sites to power rock and concrete breaking equipment, winches, drilling equipment, pumps, small tools, concreting tools, and to provide ventilation and pressurised working areas.

Depending upon the pressure and quantity of compressed air required, equipment can be obtained ranging from small portable compressors to large semi-permanent plant.

Principles of compressing air

According to Boyle's Law, the pressure (p) of a gas varies inversely with its volume (v) provided the temperature is kept constant, i.e. $p \times v =$ constant. Thus for a given mass of gas of absolute pressure p_1 and volume v_1, if changed to p_2 and v_2 the following relationship holds:

$$p_1 \times v_1 = p_2 \times v_2$$

This process of expansion or compression of a gas is called 'isothermal' and is assumed to take place without any change of temperature. However, if no heat is allowed to enter or leave the gas during expansion or compression, thereby allowing a temperature change to occur, the process is called 'adiabatic' ('isentropic'). Since higher compression is obtained under isothermal conditions (Fig. 2.1) every effort should be made to remove the heat produced by the process. In practice this cannot be fully achieved, and the production of compressed air will fall between isothermal and isentropic. This is called a polytropic process, which follows the law

$$pv^n = \text{constant (k)}$$

in which the effect of temperature is provided for by n, where

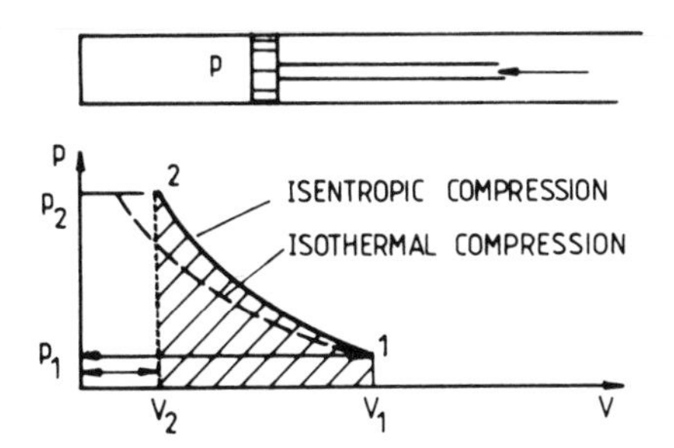

Fig. 2.1 Isothermal and isentropic pressure-volume diagrams

$$n = 1 \quad \text{for isothermal compression}$$

and $n \triangleq 1.4$ for isentropic compression.

Action in a compressor

The usual type of compressor used in construction work operates on the displacement principle, but where large quantities of air and constant flow are required, a dynamic compressor is often favoured. In the displacement type, the pressure rise is obtained by enclosing a volume of gas in a confined space, with subsequent reduction of its volume by mechanical action. A reciprocating piston, rotary screw or rotary vane provides this form of compression. In a dynamic compressor, compression of the gas is obtained by imparting kinetic energy. A centrifugal pump falls into this category.

Displacement principle of compressing air

The principles of the displacement type of air compressor may be followed by reference to Fig. 2.2. A piston (d) is driven up and down in a cylinder (a). On the downward stroke, valve (1) is open and air enters

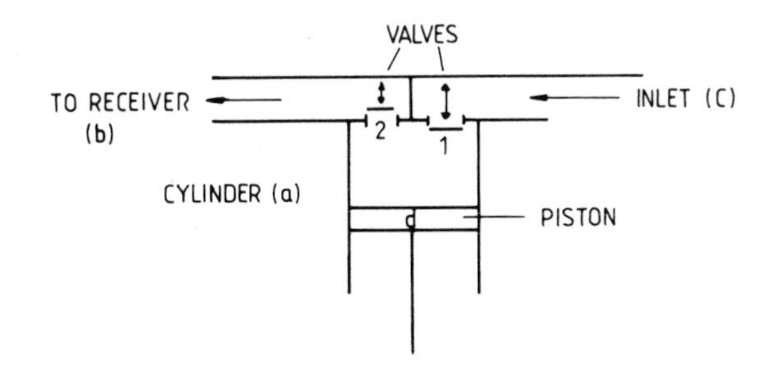

Fig. 2.2 Principles of the displacement-type compressor.

the cylinder through the inlet (c). During the upward stroke valve (1) is closed and the air compressed. The discharge valve (2) opens when the pressure inside the cylinder reaches that of the receiver (b), the receiver being a large vessel for storing the compressed air.

The process is maintained under as near isothermal conditions as possible, by removing the heat produced with a surrounding jacket of circulating water or similar.

The pressure-volume diagram of the process is shown in Fig. 2.3. At the bottom of the stroke, the volume of air in the cylinder is v_1, at atmospheric pressure p_1. As the piston moves upwards the air is compressed until the pressure p_2 of the air in the receiver is obtained, when the volume is v_2. The air is then delivered at constant pressure p_2 into the receiver and the piston continues until it reaches the top of the stroke. A small quantity (v_3) of air at pressure p_2 represents the clearance volume between piston and cover, which expands to v_4 on the downward stroke, finally reaching atmospheric pressure when the suction valve opens to admit a fresh supply of air. The cycle is then repeated, until the desired pressure in the receiver is obtained. The total work done on the air during compression, delivery and recharging is represented by area abcda.

The work done (w) can be represented by

$$dw = -v\delta p$$

but $$pv^n = \mathrm{k}$$

and for isothermal conditions $n = 1$. Neglecting the work done by the small quantity of compressed air in the clearance volume (Fig. 2.4),

$$dw = -\int_{p_1}^{p_2} \frac{\mathrm{k}}{p}\,\delta p$$

work done $$w = -\mathrm{k}\log_e p + \text{constant (c)}$$

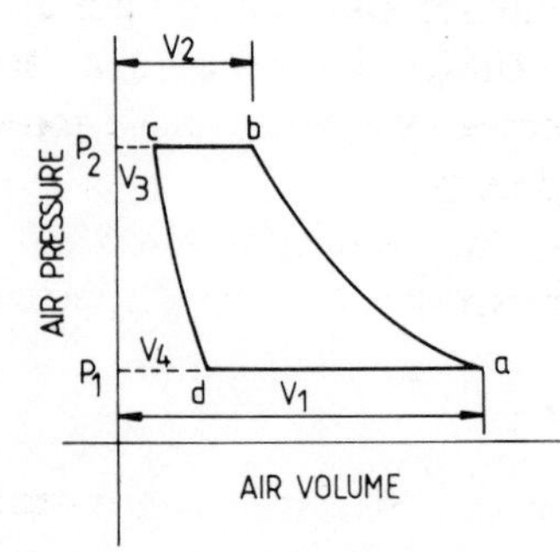

Fig. 2.3 Single stage pressure-volume diagram.

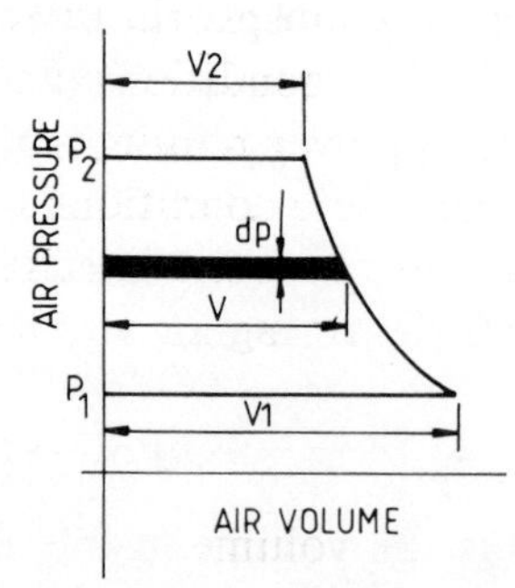

Fig. 2.4 Single stage pressure-volume diagram without a clearance volume

when $p = p_1$
$w = 0$

thus $c = k \log_e p_1$

when $p = p_2$
$w = -k \log_e p_2 + k \log_e p_1$

therefore $$w = -k \log_e \frac{p_2}{p_1} \tag{2.1}$$

but $$k = pv = p_1 v_1 \tag{2.2}$$

if $p_1 = 1.01$ bar ($0.101\ \text{N/mm}^2$) i.e. atmospheric pressure,
and $v_1 = \text{m}^3$ per min., and disregarding the minus sign.

The power required to compress v_1 from atmospheric pressure to p_2 is

$$\text{Power} = 10^6 \times 0.101 \times v_1 \log_e \frac{p_2}{0.101} \text{ Nm per min.}$$

Thus $$\text{Power} = 10^3 \times 0.101 \times \frac{v_1}{60} \log_e \frac{p_2}{0.101} \text{ kW} \tag{2.3}$$

(1 kilowatt = 1000 Nm per s)

Example

To compress 2.8 m^3 (100 ft^3) of free air per minute from atmospheric (0.101 N/mm^2) to 0.7 N/mm^2 (100 p.s.i.) indicated on the gauge (i.e. 8.01 bar absolute) requires a compressor with a theoretical power value:

$$\text{Theoretical power} = 10^3 \times 0.101 \times \frac{2.8}{60} \log_e \frac{0.801}{0.101}$$

$$\underline{9.7 \text{ kW or } 13.1 \text{ HP}}$$

Note: Air at atmospheric pressure is usually referred to as 'free air'. In practice the conditions will not be entirely isothermal and, for example, the power requirement would theoretically increase to about 13 kW if isentropic conditions were assumed ($n \triangleq 1.4$).

Taking into account mechanical losses, the required power of a compressor delivering air at 7 bar above atmospheric (gauge) is approximately:

$$\text{Power (kW)} = 6 \times v_1$$

where v_1 is the volume of free air in m^3/min, drawn in for compressing.

Multi-staging

To operate as near isothermal as possible, the compressing of air to 7 bar (gauge) (typical for powering construction equipment) is performed in two stages (Fig. 2.5). After stage 1 the air is immediately cooled before entering the stage 2 compressor. The pressure volume diagram is shown in Fig. 2.6 where the shaded area represents the work saved by two-staging. For even higher pressure rises, additional compressing stages are required.

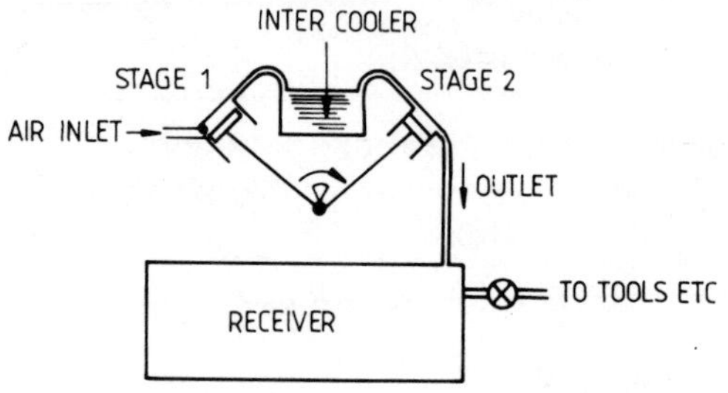

Fig. 2.5 Two-stage displacement-type compressing.

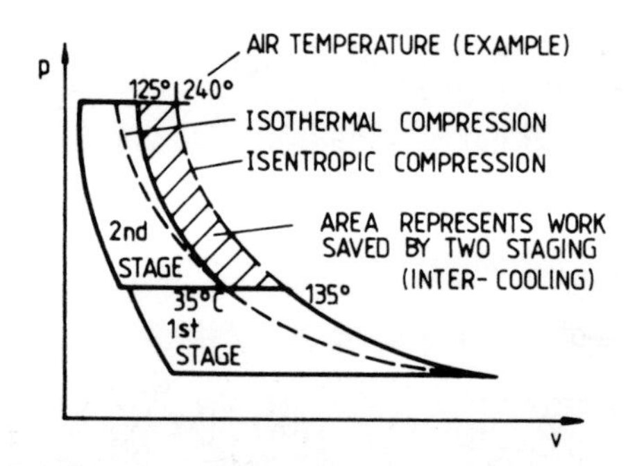

Fig. 2.6 Two-stage pressure-volume diagram.

Influence of altitude

The density of air decreases with increasing altitude and thus for a compressor operated above sea level, k in eqn. (2.2) should be reduced. For example, to compress 2.8 m^3 (100 ft^3) of air per minute from 0.05 N/mm^2 (atmospheric pressure) to 0.801 N/mm^2 (absolute pressure) requires a compressor with a power value of

$$10^3 \times 0.05 \times \frac{2.8}{60} \log_e \frac{0.801}{0.05} = \underline{6.47 \text{ kW}}$$

where $k = p_1 v_1 = 0.05 \times 2.8$
$p_1 = 0.05$

$$p_2 = 0.801$$
$$v_1 = 2.8.$$

However, the efficiency of the diesel motor to produce the power unfortunately decreases with height and therefore counteracts the advantages. Table 2.1 illustrates typical values.

Table 2.1 Effect of altitude on compressed air production

Altitude (m)	Reduction in power needed (%)	Loss of power in motor Diesel (%)	Loss of power in motor Electric (%)
1000	5	6	0
2000	10	15	8

Displacement compressor types

(i) Reciprocating compressor

A typical arrangement is illustrated in Fig. 2.2 as described earlier. Double staging can be obtained by passing air from stage 1 to stage 2 compressed through a cooler as shown in Fig. 2.5.

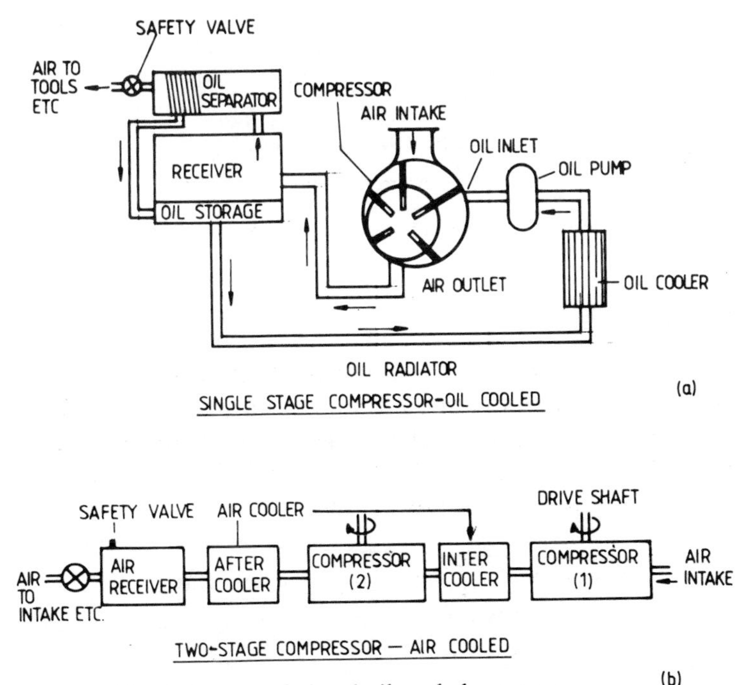

Fig. 2.7 Air cooled and oil cooled compressors.

(ii) Screw feed compressor (Fig. 2.8)

This is a new development comprising two counter-rotating screw rotor elements. The rotors do not touch and therefore lubrication is not required within the compression cylinder. Unbalanced mechanical forces and inlet and outlet valves are avoided and therefore high rates of rotation can be achieved.

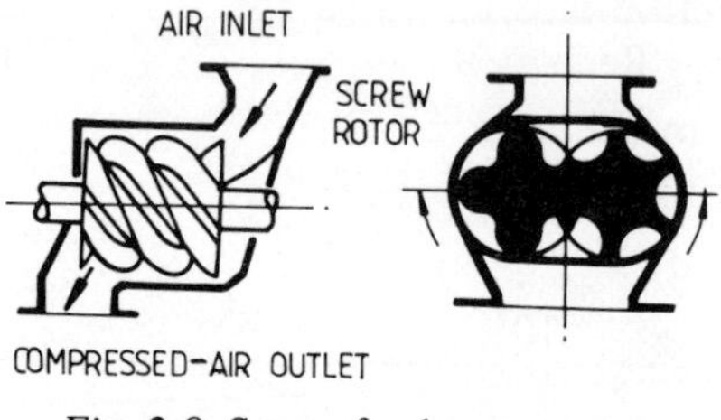

Fig. 2.8 Screw feed compressor.

(iii) Vane compressor (Fig. 2.9)

Comprises a rotor with radially adjusting blades mounted eccentrically in the cylinder. Air on entering is trapped in the space between the vanes and the cylinder wall, and as the rotation takes place (at 1500 to 1800 r.p.m.) the air volume decreases until the discharge port to the receiver is reached.

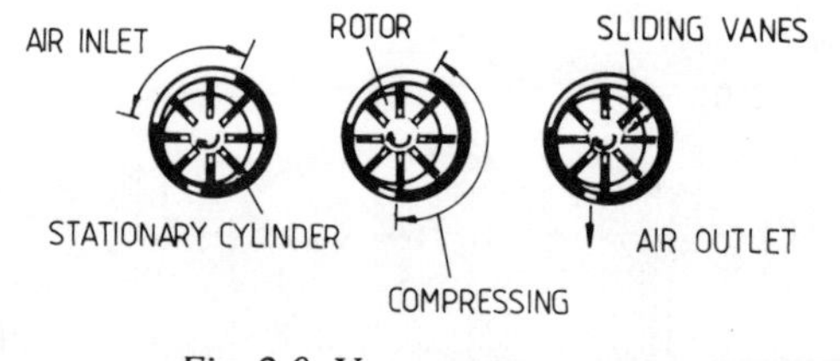

Fig. 2.9 Vane compressor.

Cooling is usually obtained by injecting small amounts of oil into the compression space, which is subsequently separated from the air and passed through a cooler, before being reintroduced (Fig. 2.7). Up to 8 bar pressure is commonly obtained in a single stage.

Regulating a compressor

The demand on the compressor depends upon the size and number of tools and equipment drawing from the receiver. The rate of production of compressed air can be regulated by varying the speed of the motor driving the compressor, whereby the throttle is automatically regulated by the air pressure in the receiver. This method is often used with electric- or turbine-driven compressors, but can also be applied to diesel engines. For most diesel-powered compressors, however, the production of compressed air is regulated by the simple on-off principle. When the pressure in the receiver reaches the maximum set, the compressor is stopped, and is only restarted when the pressure falls below a certain minimum value – the difference usually being about 10% of maximum pressure. This principle is even more basic when using the reciprocating type of compressor, where the suction valve is simply left open when maximum pressure in the receiver is obtained. The air sweeps in and out within the cylinder and thus little power is consumed.

The receiver

A receiver is required between the outlet from the compressor chamber and the connection to the tools and equipment, and serves to:

(i) Store the compressed air and so equalise air pressure variations discharged from the compressor caused by changes in demand.
(ii) Increase the cooling effect.
(iii) Assist in removal of water vapour and oil separation, where oil is the cooling fluid.

The size of the vessel depends upon the output of the compressor and the regulating method. A large vessel is required when regulation involves stopping and starting the drive motor and compressor, otherwise, when the demand varies slightly, too much starting and stopping results, causing high wear and tear on the moving parts. The approximate size of the vessel required can be determined from the following formula:

Continuous running (e.g. open valve or speed regulator) $Q \triangleq 0.1V$
Stop-start regulation $Q \triangleq 0.16V$

where Q is in m^3, and V is in m^3/min and represents the free air capacity of the compressor.

Note. A safety valve must be fitted to the receiver, set to blow off if pressure builds up by accident or malfunction.

Compressed air lines

Pressure losses are incurred in transferring the compressed air to the tools, and they depend upon:

(i) Length of pipe – increasing pipe length increases the friction losses.
(ii) Air pressure at point of entry.
(iii) Pipe diameter – the larger the diameter, the smaller the friction loss.
(iv) Rate of flow of air – greater flow increases the friction loss.
(v) Bends, fittings, valves, etc.

If the sum of these frictional resistances were to exceed the initial pressure, then there would be no air flow at all. Construction equipment is usually designed to operate at about 7 bar (100 p.s.i.) above atmospheric pressure. The efficiency of such equipment falls very rapidly as the pressure decreases, and therefore the air delivery pipe is usually designed to limit the pressure drop to within 10% of the supply pressure. The formula for estimating the drop in pressure for a particular length of pipeline is given by

$$dp = \frac{f.\,v^{1.85}}{d^5 p} \tag{2.4}$$

where dp = the pressure drop in bar
f = coefficient (82×10^3 for steel pipes)
l = pipe length (m)
d = pipe diameter (mm)
p = initial absolute pressure in bar
v = volume rate of flow (m^3/min) of free air.

For estimating purposes a chart (Fig. 2.10) can be used to determine the pressure loss for combinations of (i), (ii), (iii) and (iv) above, based on eqn. (2.2).

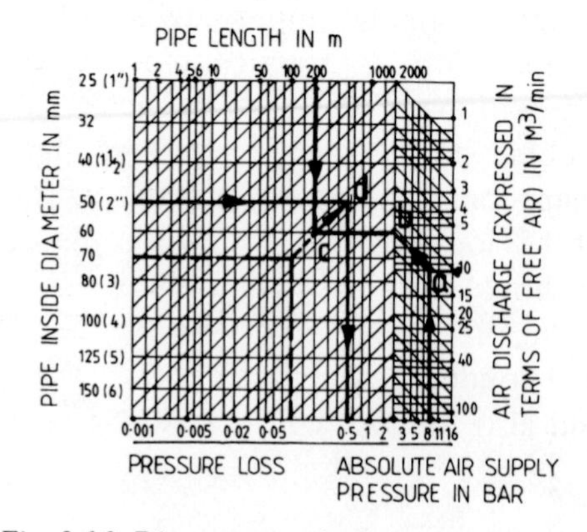

Fig. 2.10 Diagram to calculate pipe diameters

Example

Calculate the pressure loss for a 200 m length of 50 mm diameter pipe, resulting from delivering 10 m^3/min (free air) compressed to 7 bar, i.e. 8 bar absolute.

Solution
The path through the graph (Fig. 2.10) starts by entering at 10 m^3/min and proceeding horizontally to meet the vertical line drawn up from 8 bar at (a), then proceeds parallel to the sloping line at (b). This line is then projected horizontally to (c), where it meets the line drawn vertically down from 200 m. Proceed along the parallel line, to meet the line drawn horizontally from 50 mm diameter at (d). The vertical projection from (d) gives the pressure loss of 0.5 bar. An example illustrating the diagram for a 70 mm diameter pipe is shown dotted.

Other losses

Stop valve is equivalent to about 16 m of pipe friction loss.
One 90° bend is equivalent to about 5 m of pipe friction loss.
One 30° bend is equivalent to about 3 m of pipe friction loss.

These loss values should be added to the pipe length for use with the graphical method described above.

Pipe layout

It can be seen from the chart (Fig. 2.10) that pressure losses can be relatively high for small diameter pipes and therefore the pipe length should be kept to a minimum. In practice, pipes for construction application are generally within 25–150 mm diameter, but for flexibility, textile reinforced rubber hose about 25 mm diameter is mostly used in short lengths to snap connect to the tools and equipment. Where the compressor cannot be located near the work place, a long pipe of sufficient diameter should be laid down and flexible connections made to it. Such pipes when laid outdoors should preferably be buried to avoid freezing of the condensate in winter (otherwise ice may form and cause blockage), and also for protection against damage caused by site traffic.

Compressor rating

Compressors are rated in terms of the volume of free air per minute taken in for compressing. They are available as portable, towed or permanent units with capacities ranging from 1 m^3/min to more than 100 m^3/min. It can be seen from eqn. (2.3) that an increase in the output pressure requires additional power to compress a given quantity of free air. Thus the free air rating is usually stated at a particular delivery pressure, e.g. 7 bar (gauge). Two stage compression through an intercooler is commonly used to produce air at 7 bar gauge pressure to minimise the adiabatic effects.

Typical air consumptions by construction equipment

	Free air consumed (m^3/min) *delivered at 7 bar gauge pressure*
Heavy concrete breaker	2.5
Medium concrete breaker	1.5
Light concrete breaker	1.0
Clay digger (spade type)	0.9-1.25
Picks	1-1.2
Hand held sinker drills	2-4
Feedleg drills	4-6
Rig mounted rotary and rotary-percussive drills	5-10
Drill hole flushing	Up to 5
Vibrators	1-2
Small tools	0.5-1
Hoists and winches	Up to 10
Pumps	2-5
Pile hammers	1-50

Note: Higher working pressures may be necessary when operated in compressed air chambers, e.g. tunnelling.

References

Atlas Copco Manual – Atlas Copco AB, Stockholm, Sweden.
Various literature from:
Broom & Wade Ltd., High Wycombe, Bucks
Compair Ltd., Camborne, Cornwall
Gardner–Denver International, Quincy, U.S.A.
Ingersoll–Rand, Phillipsburg, N.J., U.S.A.

CHAPTER THREE

Rock drilling

Rock boring is required on a variety of civil engineering projects including:

(i) core sampling for geological investigations
(ii) confining blasting charges in quarrying and tunnelling
(iii) rock bolting and anchoring
(iv) grouting.

There are basically three methods of producing holes in rock, each suited to a particular application. These are:

(i) rotary drilling
(ii) rotary-percussive drilling
(iii) percussive drilling.

A fourth method, involving intense heat concentrated on a confined part of the rock, is occasionally used. Other developments are progressing with laser beams and high pressure water jetting. Such methods are at present only in the pilot stage and it may be some years before they are sufficiently reliable to be adopted by the construction industry.

Choice of method

For many construction applications, such as in quarrying, rock bolting, grouting and tunnelling, where relatively shallow holes less than 50 metres deep and up to 100–150 mm diameter are required, rotary-percussive equipment provides a light, manoeuvrable and efficient method in medium to very hard rock. For larger diameters or when boring to greater depths and accuracy, or where soil or soft rock is to be encountered, rotary drilling machinery is required. Percussive drilling only is very slow, but because it is a simple and basic method, it is sometimes more convenient to use this technique in soils investigation for breaking through a minor rock obstruction. The merits of the various methods are shown in Table 3.1 and the classification of the rocks appropriate for each is given in Table 1.2 on page 5.

Table 3.1 Comparison of drilling methods

Method	Rock type				Normal application	
	Soft	Medium	Hard	Very hard	Max. borehole dia. (mm)	Max. borehole depth (m)
Percussive	****	****	****	****	400	No limit
Rotary-percussive						
Drifter drilling	*	**	***	****	150	40
Down hole drilling	***	****	****	****	200‡	250
Rotary						
Cutting	****	****	*	—	600	5000
Crushing	***	****	****	***	600§	5000
Abrasive (diamond drilling)	—	**	****	****	150	2000†
Thermal	—	*	****	****	150	5

* = poor **** = good
† For core sampling, otherwise as for crushing method
‡ Bits are available up to 800 mm diameter for special applications
§ Roller bits up to 6 m diameter are available for special tasks

Rotary drilling

Rotary drilling (Fig. 3.1) relies on a high feed thrust applied down the drill stem, to force the edges of the bit into the rock surface. High torque and rotation of the drill shaft then cause cracking and chipping, and rock fragments are broken away. The rotary drill is supported on a mast above the hole, and additional drill rod extensions are inserted into the rotor to deepen the borehole. Either a compressed air or more usually a hydraulic or an electric motor is used to power the unit.

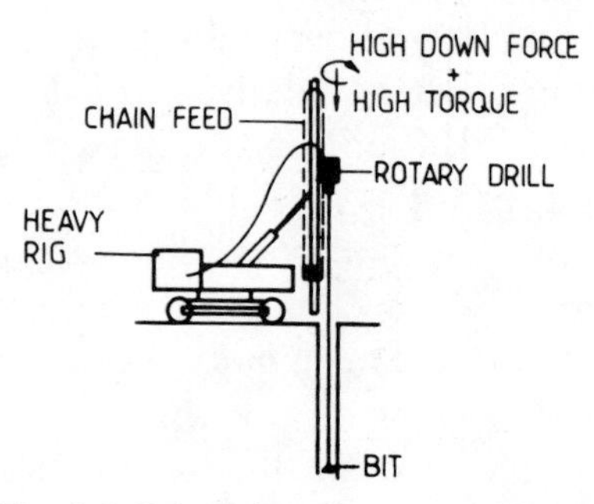

Fig. 3.1 Principles of rotary drilling.

Flushing of the drill hole may be carried out with water or compressed air. For versatility in construction work, rigs are generally truck-mounted.

Rotary drilling methods

Rotary cutting (Fig. 3.2)

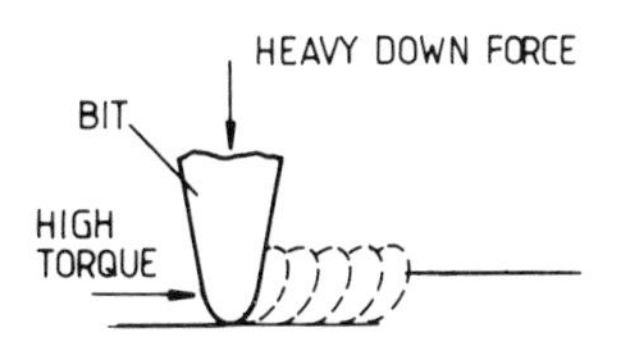

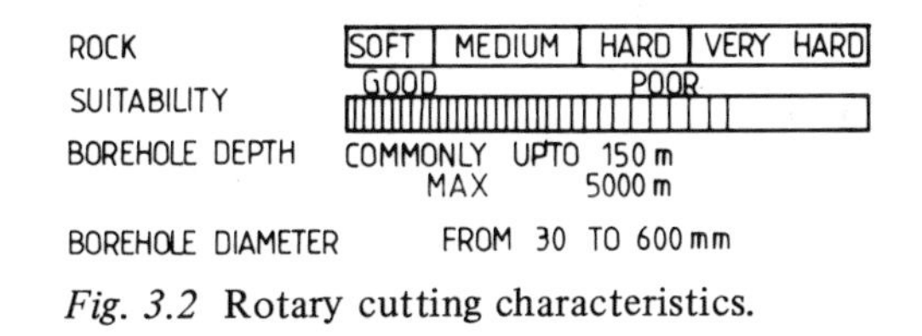

Fig. 3.2 Rotary cutting characteristics.

Winged bits (Fig. 3.3) with tungsten carbide inserts, to aid penetration and durability, are suitable for holes up to about 150–200 mm diameter. For wider borings it is common practice to follow the primary bit with staged reamers (Fig. 3.8).

Current bit design limits the method to applications in soft rock and requires a feed force up to 0.5 kN/mm bit diameter to facilitate

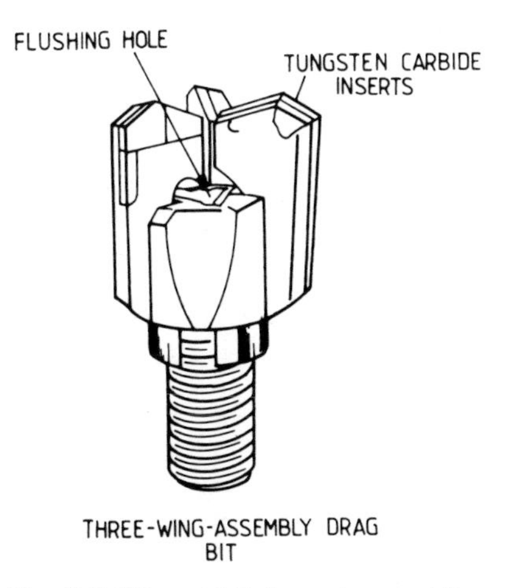

Fig. 3.3 Winged bit for rotary cutting.

adequate penetration of the inserts. Thus a 200 mm diameter bit in medium hard rock would require a rig capable of delivering 10 tonnes (100 kN) of feed force. The torque and feed force required to sheer the rock plane depends upon the borehole diameter and rock strength. The bit is rotated within the range 50–150 r.p.m.: the harder the rock, the slower the rotation speed.

Rotary crushing (Fig. 3.4)

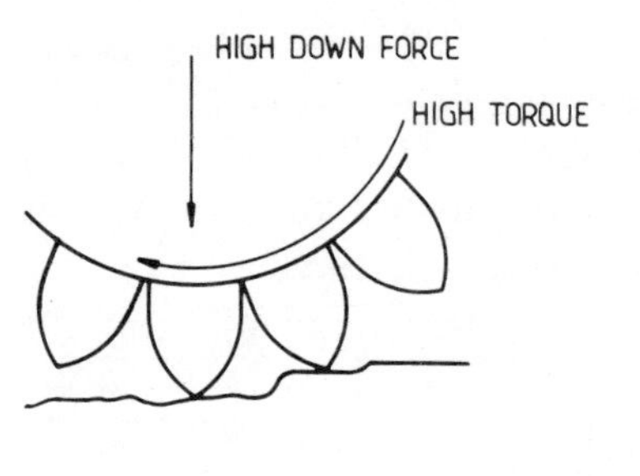

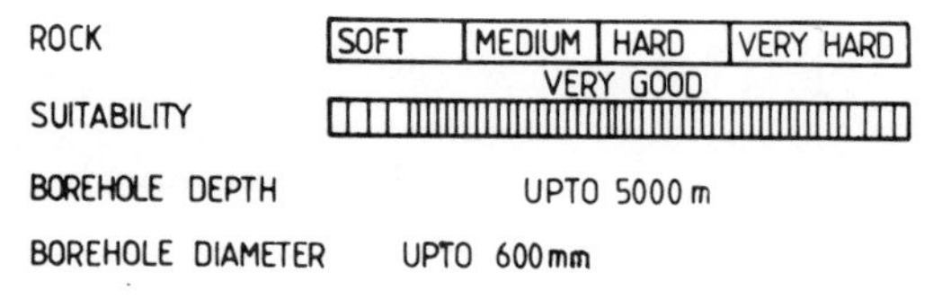

Fig. 3.4 Rotary crushing characteristics.

Toothed and button roller bits

During the last 10–15 years development of roller bits by the oil industry has been updated for more general application(s) by construction users and today they are commonly chosen for medium to hard rock. Like the rotary cutting method, a high feed force must be applied to the bit, but instead of a cutting edge, steel cones with tungsten carbide tipped teeth (Fig. 3.5) or buttons (Fig. 3.6) rotate on their own axis and the rock is crushed and chipped away. Because the drill stem is also rotated, a fresh part of the hole base is always worked on. The button-type bit is favoured for harder rocks.

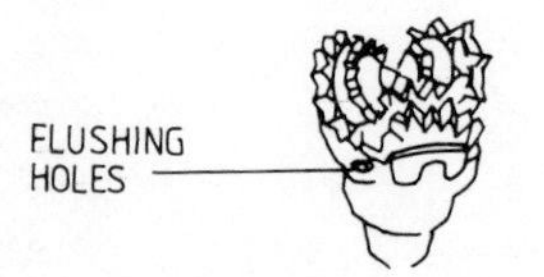

Fig. 3.5 Rotary crushing bit.

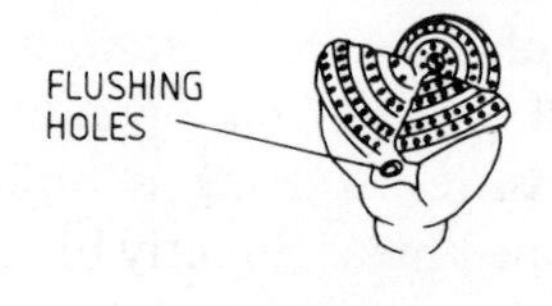

Fig. 3.6 Roller cone bit.

The roller bit method relies on overcoming the compressive strength of the rock, and high feed pressures up to 1.5 kN/mm of bit diameter over the borehole cross-section may be required in hard rock. Thus for a 200 mm diameter borehole a feed force of 30 tonnes (300 kN) may be necessary and 75 tonnes for a hole of 500 mm diameter bore. Bits, however, are available in smaller sizes down to 75 mm, requiring less feed force. Only light truck-mounted rigs are then needed.

Effect of feed force

The penetration rate is directly dependent upon the feed force applied to the drill bit. However, as the force is increased there is an optimum at which the resistance to turning of the bit exceeds the advantages gained from the extra down-force, as shown in Fig. 3.7.

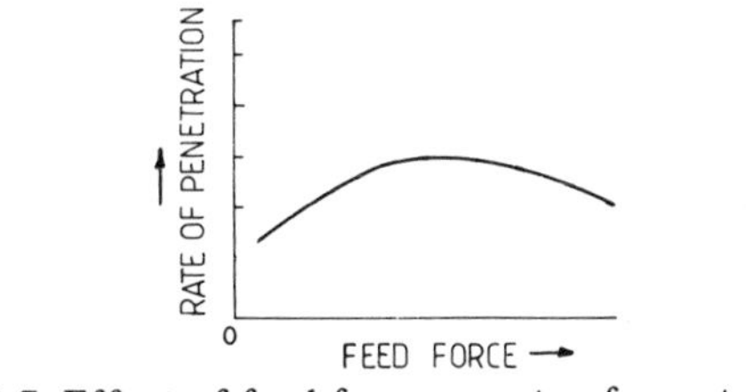

Fig. 3.7 Effect of feed force on rate of penetration.

Factors limiting the depth of drilling

The limit to the depth of hole that can be bored by rotary methods is largely governed by the quality of drill bit, the bit life (rods must be withdrawn to change the bit), strength of the drill rod, the capability of the feed equipment (as relatively high torque and down feed forces are required) and the flushing efficiency. Also, a long drill stem is heavy and so provision of sufficient hoisting capability is another important limiting factor.

Curved boreholes

Recent developments in drilling technology now make it possible to deflect the borehole from the vertical (Fig. 3.8). A typical method involves a hydraulic rotary motor which is located at the bottom of the drill string, and operated by drilling mud fed down the stem under pressure. Steering is obtained by inserting a specially designed wedge-shaped piece directly behind the motor. The drill bit is attached directly to the motor, and techniques such as mechanical plumbs, electronic dipmeters, acid-etch, photography, etc. are used to measure the

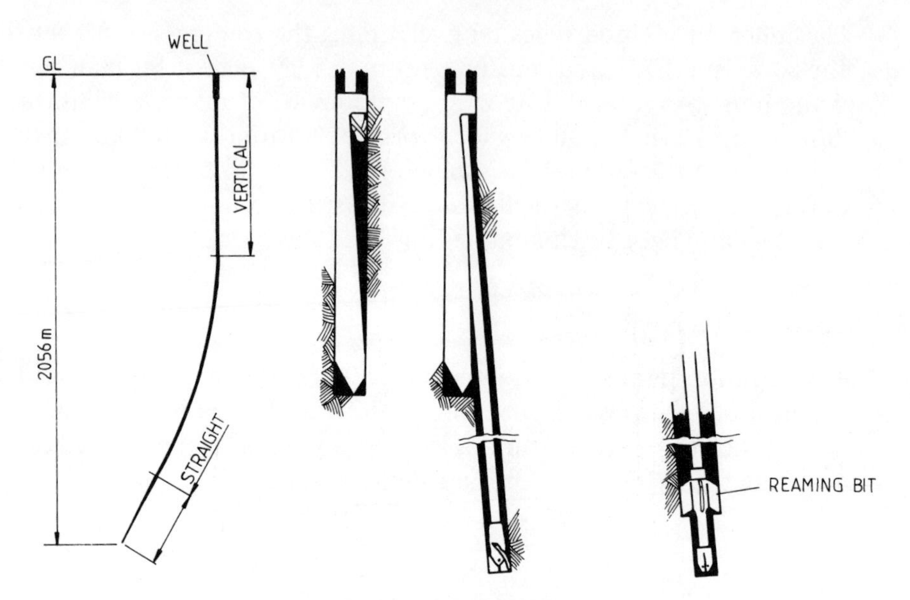

Fig. 3.8 Deflecting a drill hole from vertical.

deflection so that adjustments can be made to the directional path.

Rotary drilling rigs

The drilling rig (Fig. 3.9) comprises:

(i) Rotary drill (usually hydraulically driven) with variable revolution rate control.
(ii) Hydraulic or chain system to provide down-thrust to the drill stem and bit.
(iii) Suitable support mast to raise and support the drill stem.
(iv) Swivel barrel loader to position the drill stem insertions.
(v) Winch to raise the drill stem and bit.
(vi) Centralising chuck to hold the stem and casing on line when drilling.
(vii) A vehicle to support the engine, drilling equipment, winches, compressor and flushing pump.

Typical data for drilling rigs are given in Table 3.3.

Currently, rigs can be equipped to produce boreholes 5000–6000 m deep. However, for most construction applications, medium to light rigs, truck-, wheel- or crawler-mounted, are usual, with the larger vehicles being suitable for boreholes up to 200 mm diameter and for drilling to depths of up to 100–200 m. The truck-mounted version is

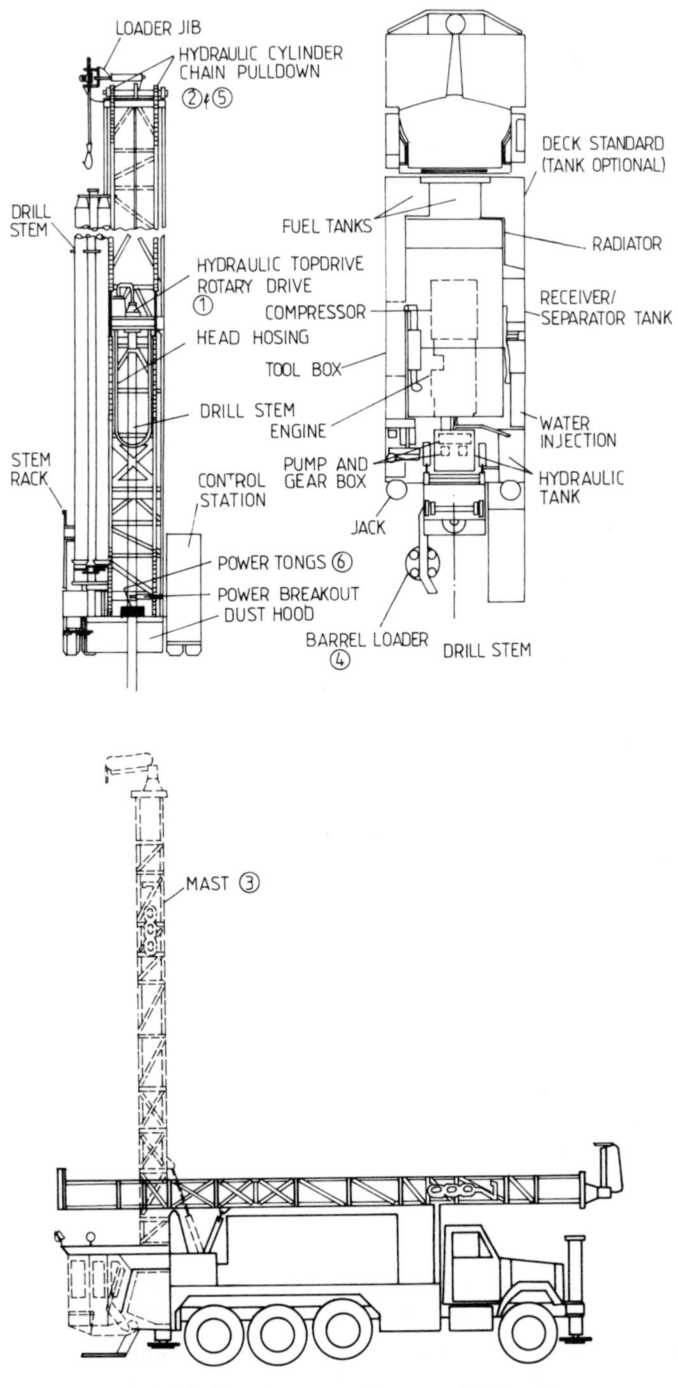

Fig. 3.9 Truck-mounted rotary drilling rig.

Table 3.3 Rotary drilling rigs

	Rig for boreholes	
	up to 125 mm diameter	up to 250 mm diameter
Torque	6 kN.m	10-12 kN.m
Feed force	50 kN	180 kN
Hoist capacity	50 kN	100 kN
Rotary speed	25-150 r.p.m.	25-150 r.p.m.
Rotary motor power	30 HP (22 kW)	100 HP (75 kW)
Hoist speed	30 m/min fast, 6 m/min slow	5-30 m/min
Feed speed	0-5 m/min	2-16 m/min
Mast	Up to 7 m stems	Up to 9 m stems
Angle drilling	0-45°	0-30°
Total weight		Approx. 28 tonnes

Note
(1) Air flushing – up to 20 m^3/min
Mud or water flushing – up to 600 l/min
(2) Rotary drills and rigs are often suitable for use with 'down the hole' drills as a rotary-percussive method.

particularly useful, because of the advantage of its mobility. Most of the modern rotary rigs can be quickly converted to deal with soils and different rock hardnesses, simply by changing the bit or even switching to the rotary-percussive technique.

Table 3.4 Production data for rotary drilling performance –200 mm diameter borehole

Rock classification	Bit loading (kN/mm bit diameter)	Penetration rate (m/h)	Bit life (m of hole)
Soft	0.3-0.6	20-45	1500-6000
Medium	0.5-0.8	10-20	600-1500
Hard	0.7-1.0	5-10	150-6000
Extremely hard	0.9-1.3	2-10	50-150

Diamond drilling (rotary abrasion) *(Fig. 3.10)*

Diamond drilling is mostly used in mineral exploration where core samples of the rock are required for analysis. The method is suitable for borings of up to about 150 mm diameter.

Drilling takes place on a similar principle to other rotary methods, and requires very high feed force, but relatively lower torque than

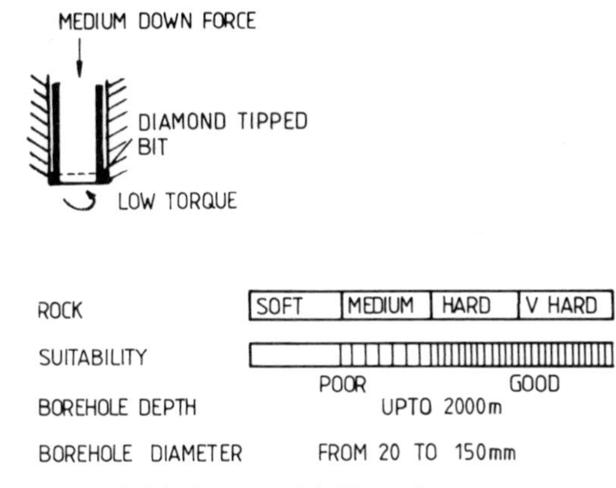

Fig. 3.10 Diamond drilling characteristics.

rotary drilling. Borehole depths of over 2000 m are not unusual with rotary speeds of up to 1000 r.p.m. The method relies on abrasion, using a diamond-impregnated or surface-set bit.

Diamond bits *(Fig. 3.11)*

Method of action of diamond bit

The general opinion is that the combined pressure and rotation applied to a single diamond causes plastic deformation of the rock. In hard, brittle rock the compressive stresses break the material along the zone of maximum sheer stress and the small fragments must then be continuously and efficiently removed by a flushing fluid to prevent obstruction.

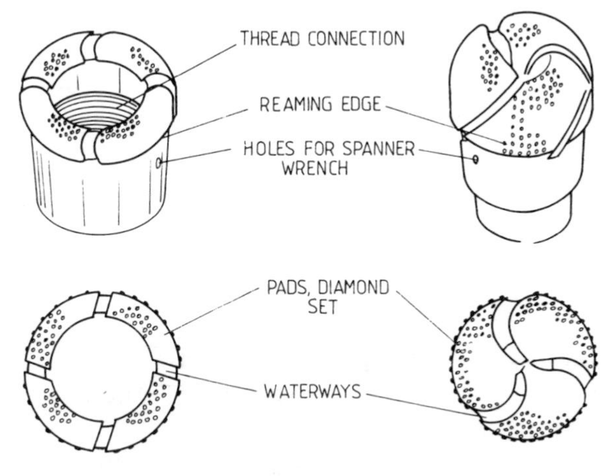

Fig. 3.11 Diamond bits.

Thus the rate of penetration of the bit is influenced by the rotation speed, feed pressure and pumping rate of the flushing medium.

Rotation

For diamond surface-set bits, rotary speeds of 50–150 m/min are used; slightly higher speeds of 100–200 m/min are possible with impregnated bits. Thus for a 50 mm diameter bit, 150 m/min represents approximately 1000 r.p.m. It has been shown by Marx that for a constant feed pressure, the theoretical rate of penetration is proportional to the rotation speed of the bit. Thus, as the diameter of the bit is reduced, the revs per minute must be increased to maintain an adequate rotation speed at the radius of the diamond surface. However, the penetration rates also depend upon the rock hardness, for example a rotation speed of 120 to 180 m/min is suitable in soft formations, whereas 60 to 120 m/min is preferred for a hard formation. The rates of penetration may not, of course, be the same for both.

Feed force

Each diamond in the bit must transmit a stress which exceeds the strength of the rock, and clearly, therefore, there is a close relationship between bit load and penetration rate.

Thus

$$G \leqslant \frac{P}{d \times a} \leqslant D$$

where D = diamond strength
G = rock strength
P = bit load
d = number of diamonds
a = contact surface area of diamond

The following loads per carat are recommended for different rock types:

Rock type	*Load per carat (N)*
Granite	200–1000
Quartz	400–2000
Basalt	400–1800
Basalt lava	100–600
Sandstone	200–1000
Shale	200–400
Limestone	20–1200

Thus for a given size of diamond, i.e. stones per carat and density of stones in the cutting surface, the required feed load on the bit can be determined.

In soft to medium hard rock, diamonds of 20 to 90 stones per carat are selected, while for hard to very hard rock 200 to 250 stones per carat are necessary, i.e. the harder the rock, the smaller the diamond and greater the number of diamond stones.

Diamond loss and wear

The edges of active diamonds get rounded as they are subject to harsh erosion forces. This is compounded by high rates of vibration, which cause some shattering and sheering of the diamonds. Insufficient cooling by the flushing fluid may also cause burring of the diamond. The carat loss is related to the length of borehole drilled by the bit, as shown in the approximate data given in Table 3.5.

Table 3.5 Diamond loss and bit life

Rock type	Carat loss per m of penetration	Bit life (m)
Basalt	0.2	10–60
Dolomite	0.03	25–60
Granite	0.2	10–30
Limestone	0.05	25–150
Sandstone	0.05	25–150
Quartz	0.5	5–30

Diamond drilling equipment

The drilling equipment is similar to that required for normal rotary boring, except that the feed force is usually delivered through hydraulic rams rather than chain feed. The drill may be mounted on wheels, tracks or skids. The rig shown in Fig. 3.12 comprises a diesel engine or electric motor, skids or tracks and base frame (1), main hoist and cat head to handle the drill rods (2), hydraulically-powered rotary drill and hydraulic chuck to supply the downfeed force (3), and drill rods. Typical data to handle 900 m of drill rods of 40 mm diameter are as follows:

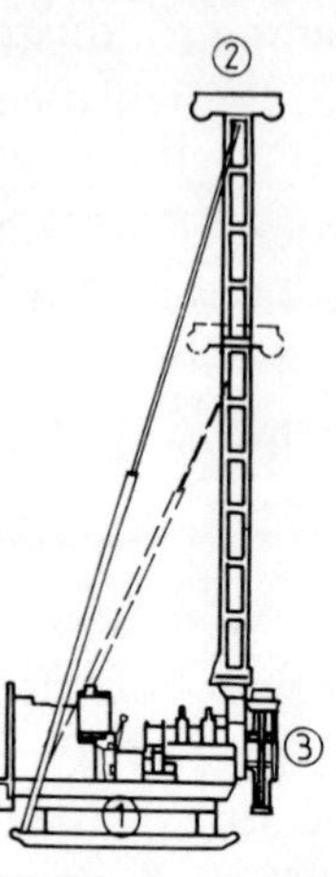

Fig. 3.12 Diamond drilling rig.

Torque	1.5 kN.m
Feed force	45 kN
Hoist capacity	60 kN
Rotary speed	100–1500 r.p.m.
Engine power	40 HP (30 kW)
Drill rod diameter	up to 45 mm
Hoist speed	30 to 150 m/min. 3 m or 6 m long drill rods
Mast	up to 8 m
Feed length	0.5 m
Angle drilling	0° to 360°
Flushing	Water or air supply are additional requirements.

It can be seen in Fig. 3.13 that the chuck and mast can be turned to accommodate angle drilling. Drill rod extensions are raised into position by the mast winch, and the chuck is simply unclamped and repositioned when its full 0.5 m range has been reached during the drilling operation. The new extension is thereby passed down and through the rotary unit as drilling proceeds.

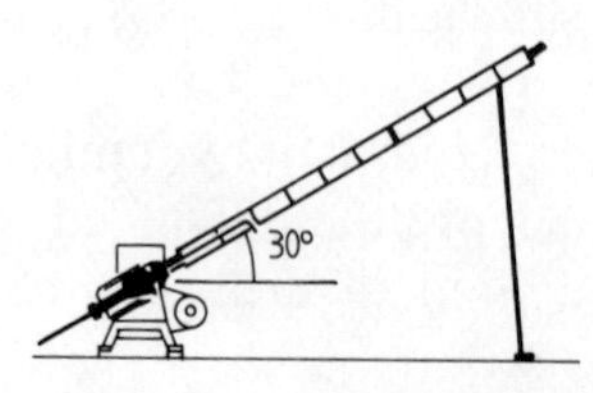

Fig. 3.13 Angle drilling.

Core sampling

Core sampling requires special tubular edged bits. In hard, compact formations the single tube core barrel is often used (Fig. 3.14). However, for less firm material the double tube core barrel is preferred (Fig. 3.15), the advantage being that the flushing water is shielded from the core and so minimises the washing action. The inner tube is normally of the 'swivel type', thus reducing the risk of breaking the core.

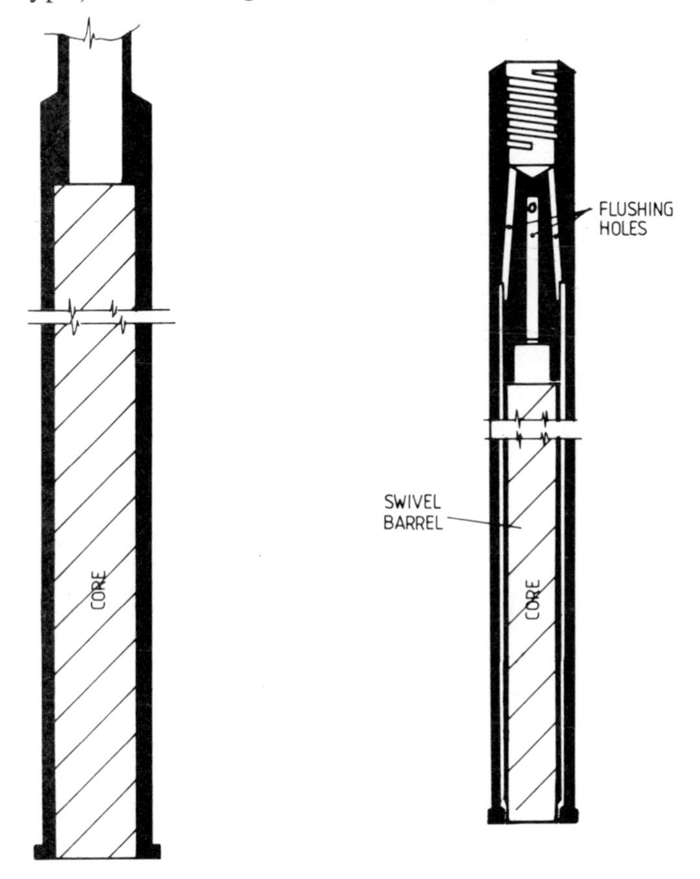

Fig. 3.14 Single tube core barrel. *Fig. 3.15* Double tube core barrel.

Unfortunately, with these methods the whole of the drill stem must be withdrawn from the borehole to remove the core sample (300–400 mm long), thus resulting in unproductive time. To overcome this problem the Wire Line core barrel was introduced, making it possible to take the core sample from the barrel without pulling the string of drill rods (Fig. 3.16).

Core samples up to 150 mm diameter can be taken with diamond bits. In soft, abrasive rock or in mixed formations where large diamonds of 8–15 stones per carat would be necessary, it is sometimes possible to use tungsten carbide particles mixed with special alloy-tipped bits instead of diamonds, as these are less expensive.

Core drilling bits have a tendency to form a slightly tapered hole, but with the insertion of a reaming shell (Fig. 3.17) the correct gauge is maintained and so a new bit can be inserted into the borehole without damage.

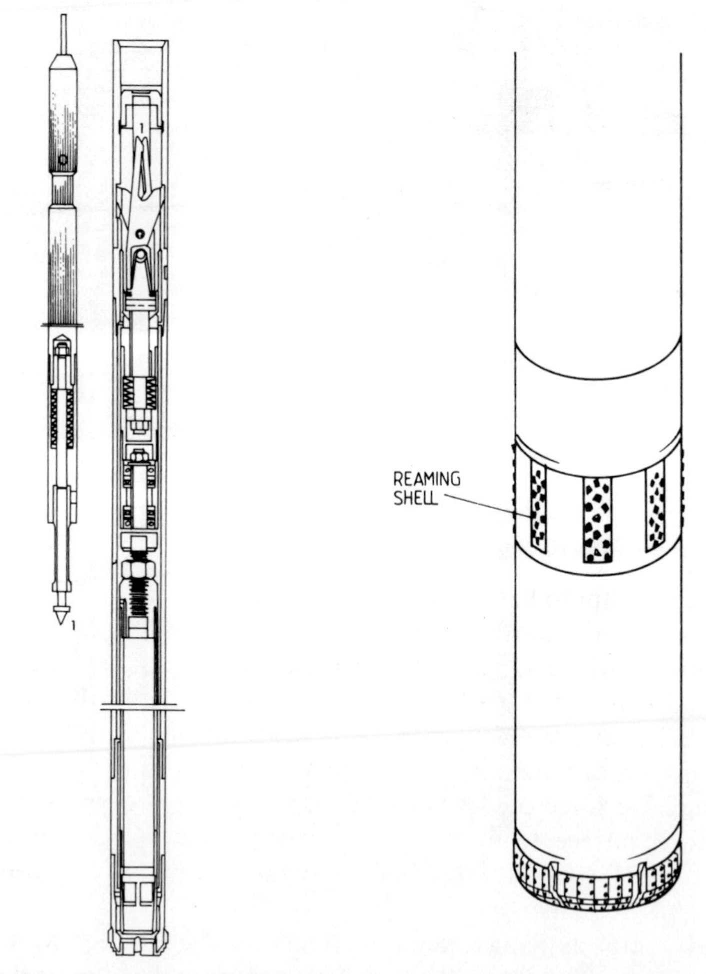

Fig. 3.16 Wire line core barrel.

Fig. 3.17 Reaming shell.

Casing shells for rotary and diamond drilling

Where the borehole must pass through unstable rock or soil, or where the flushing medium would be absorbed into the surrounding formation, casing tubes may be required. The method basically involves following the drill bit and stem with a casing bit and lining tube of a slightly larger diameter. Casing shoes (Fig. 3.18) for drilling have diamond-tipped cutting segments (Fig. 3.19) on the exterior lip, while the interior is smooth. The inner diameter must be sufficient to permit free passage of the drill bit, stem and core sample.

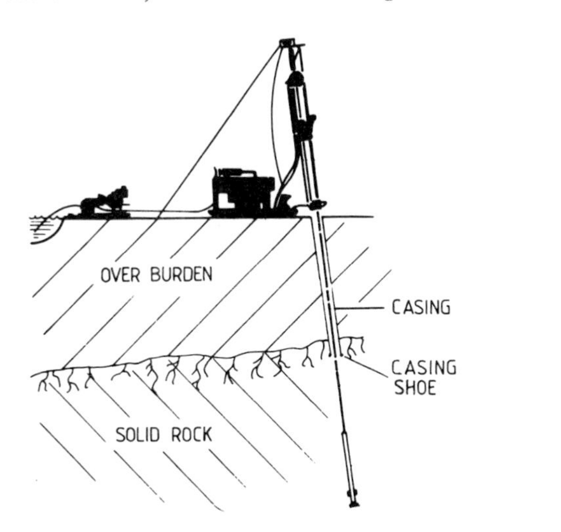

Fig. 3.18 Hole lining or casing.

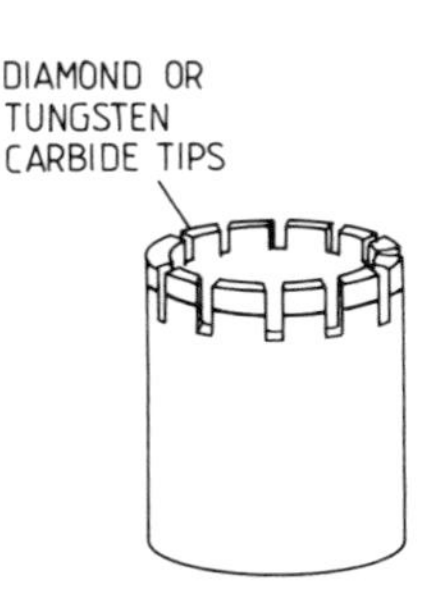

Fig. 3.19 Casing shoes.

Rotary-percussive drilling *(Fig. 3.20)*

Where medium to hard rock is to be encountered throughout the entirety of the drilling, the rotary-percussive method is often favoured because the rig is light and good rates of penetration can be obtained. The method is used for blast holes, rock anchors, grouting holes, wells, etc. In rotary-percussive drilling the drill bit is supplied with both a percussive and a rotary action rather than with a high feed force as in rotary boring. The force of the blow causes penetration of the bit inserts into the rock surface to form a crater, which is subsequently disturbed by the turning motion. The percussive action provides a considerably greater force than would be achieved with the same load applied without impact. The broken rock fragments are removed with either air or water flushing introduced under pressure down the drill stem and out through the base of the bit.

Currently the two basic methods are drifter drilling (Fig. 3.21), where the drill is located at ground level, and 'down the hole' drilling (Fig. 3.22), where the percussive part of the drill actually follows the bit down the borehole.

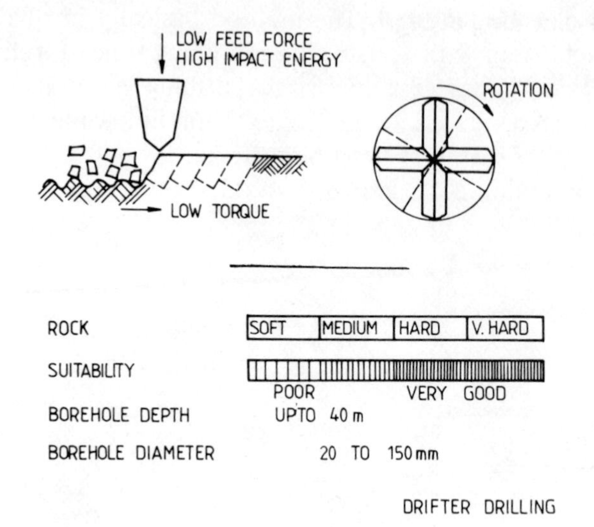

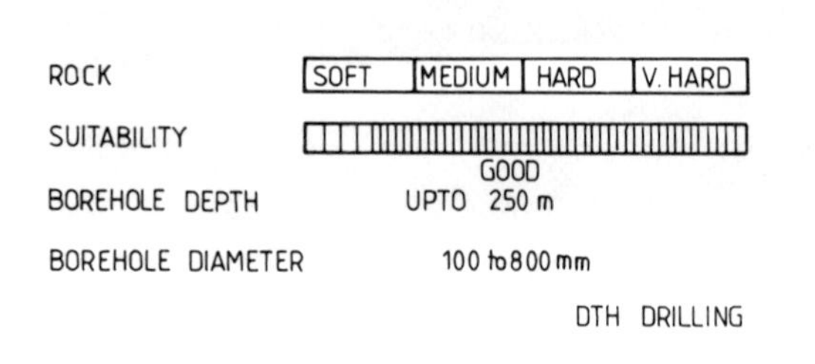

Fig. 3.20 Rotary-percussive drilling characteristics.

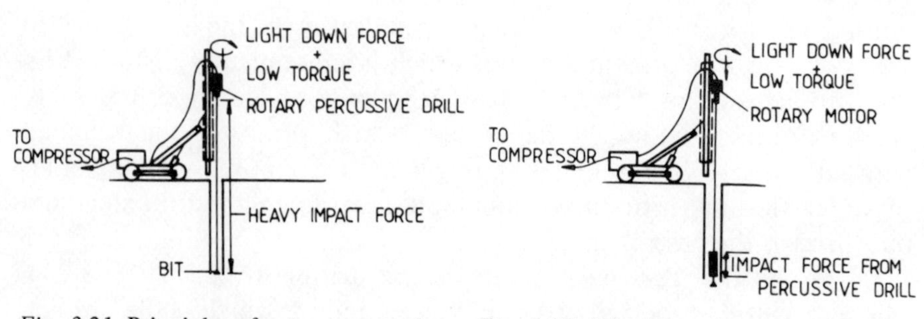

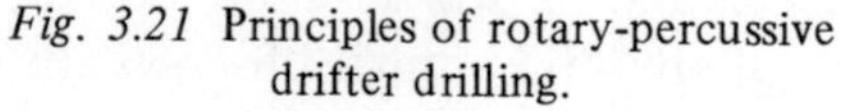
Fig. 3.21 Principles of rotary-percussive drifter drilling.

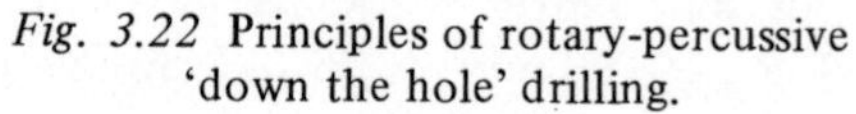
Fig. 3.22 Principles of rotary-percussive 'down the hole' drilling.

Rotary-percussive drilling equipment

The drilling rig (Fig. 3.23(a)) consists of:

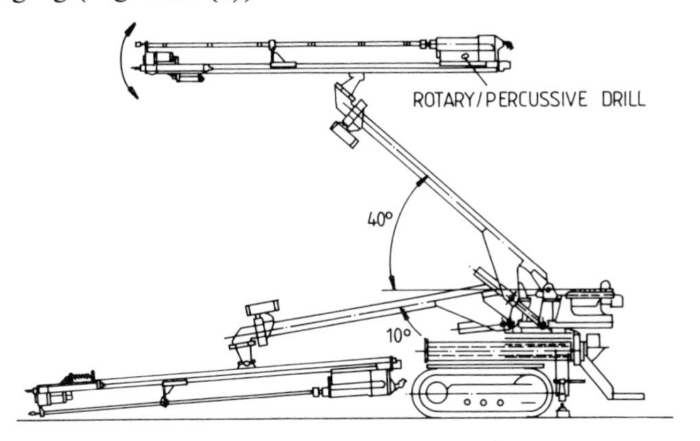

Fig. 3.23(a) Drifter drilling rig.

(i) A compressed-air driven rotary-percussive rock drill.
(ii) Chain feed to maintain the feed force on the bit.
(iii) Mast or leaders to support and guide the drill.
(iv) Centralising chuck to hold the drill rod.
(v) Telescopic boom.
(vi) Tracks or similar.
(vii) Compressed-air supply.

Tracks are usually selected for surface drilling and the boom can be tilted and repositioned to accommodate vertical, horizontal and angle drilling as shown in Fig. 3.23(b).

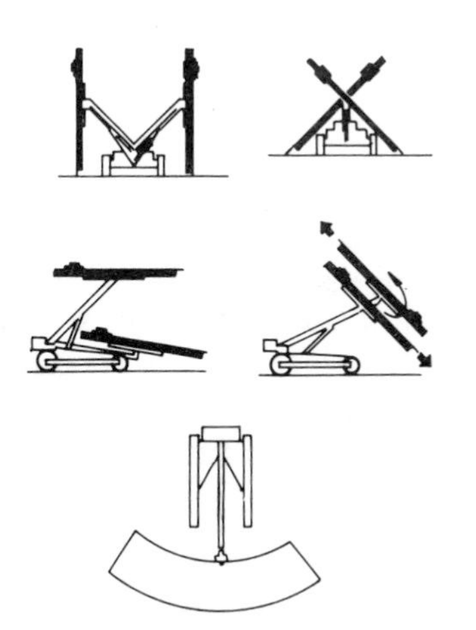

Fig. 3.23(b) Angle drilling movements.

Compressed air drifter drill *(Fig. 3.24)*

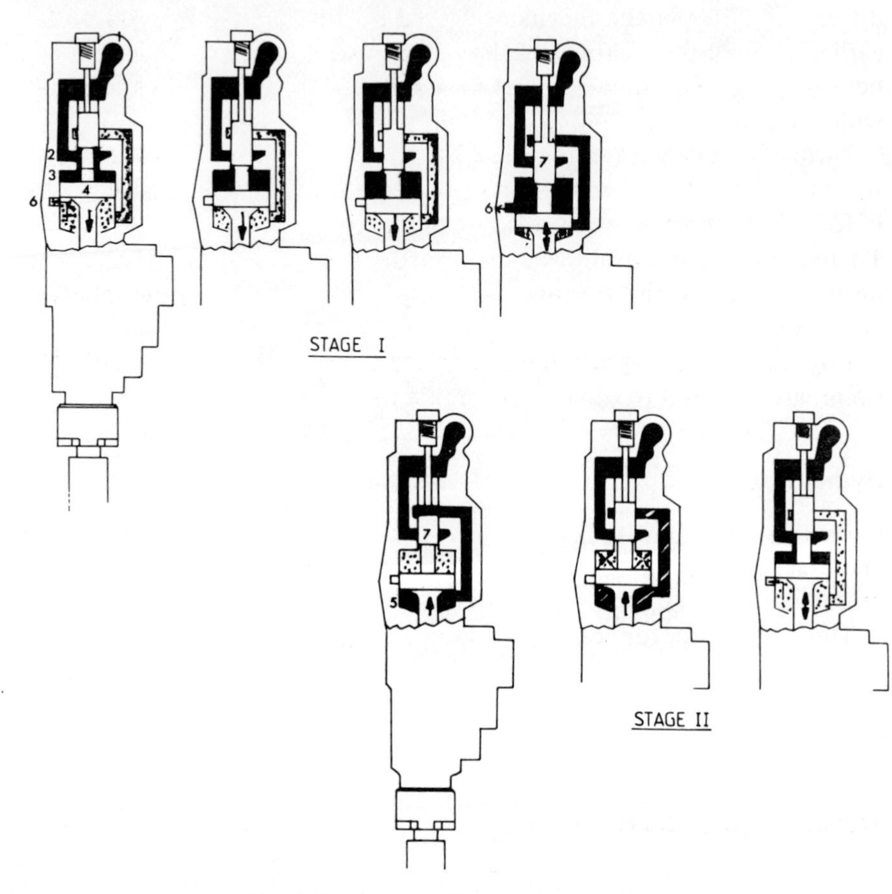

Fig. 3.24 Drifter drill working principle.

Operation of the drill is generally by means of compressed-air, whereby kinetic energy provides the percussive blow, as follows:

Stage 1. Compressed-air enters 1, through the support port 2 and into the cylinder 3, pushing the piston 4 forward. The exhaust port 6 meanwhile allows the air in the lower part of the cylinder at atmospheric pressure to be expelled until the exhaust port 6 is blocked.

Stage 2. The piston continues to move downwards under its own momentum to uncover the exhaust port 6. Meanwhile the compressed-air is shut off by the piston control head 7. During this phase the control head 7 allows compressed-air to enter the lower part of the cylinder 5, causing the piston to move upwards until the cycle is completed, ready to start again. Thus a single blow has been delivered to the striking bar.

Rotation of the drill bit is achieved by one of two basic methods, the difference being in the mechanism used for turning the drill rod. In the earlier models the piston travels up and down in the cylinder along a helical spline. The spline is part of a rifle bar which operates a ratchet which allows rotation in one direction only.

Unfortunately with this type of mechanism, when the resistance to turning of the drill is greater than the torque output, the piston stroke is forced to shorten, with a consequent loss of power and efficiency. To redress this disadvantage, modern drifter drills have a separate air motor to provide the turning effect. As a result, higher blow rates can be achieved.

For both types, approximately 1500 blows per minute at 100–150 r.p.m. are required to obtain acceptable penetration rates.

Hydraulic drifter drill

Recently, hydraulically-powered drifter drills have appeared, which manufacturers claim to be more efficient than the equivalent compressed-air drills i.e. up to twice the penetration rate.

The rotary motor can be varied from 0–300 r.p.m. Because the hydraulic percussive mechanism allows the use of a slim piston, greater energy flow through to the stem is obtained, thereby raising drilling performance. Either compressed-air or water may be used for flushing.

'Down the hole' (DTH) drill *(Fig. 3.25)*

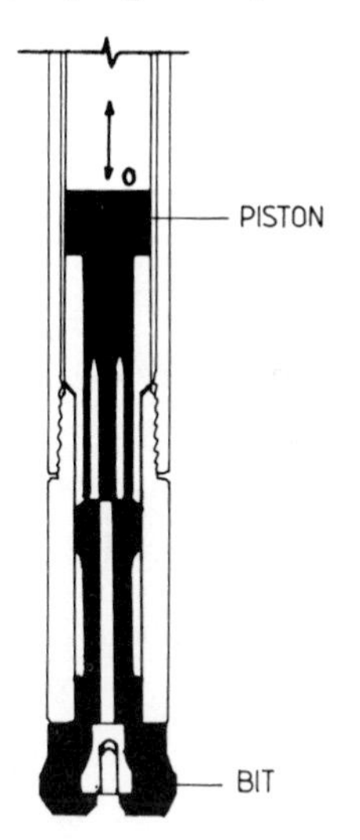

Fig. 3.25 'Down the hole' drill working principle.

The conventional drifter drill described above becomes less efficient as the length of borehole increases. This is due to loss of impact energy into the drill stem itself, and, more importantly, impact energy is dissipated as heat at each drill rod connection. Practical tests have shown that the energy loss across a rod joint is approximately 10% of the energy in the rod before the joint. Thus the rate of penetration will decrease as the depth of borehole increases.

To overcome these difficulties a 'down the hole' (DTH) drill was developed (Fig. 3.22). The rotary motor remains above ground level while the bit is followed down the hole by its pneumatic/hammer, to produce a virtually constant drilling rate. Unfortunately, because of restrictions on the practical size of piston, etc., the minimum size of borehole is about 100 mm diameter, thus restricting the available power to the drill. Thus for shallow holes of this approximate diameter, the selection of a more powerful drifter drill would allow much faster drilling rates to be obtained.

However, because of the loss of efficiency with depth of drilling when using the drifter (Fig. 3.26), a break-even point with the DTH would eventually be reached since there is virtually no deterioration in drilling rate with the DTH machine, and so for deep holes a lower-powered DTH drill would be more efficient than the larger drifter drill.

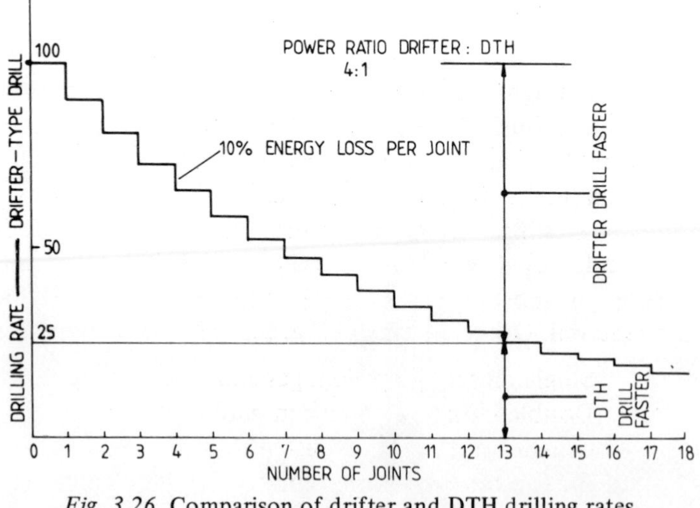

Fig. 3.26 Comparison of drifter and DTH drilling rates.

Effect of air pressure

Most drifter drills are operated with air pressure at 100–150 p.s.i. (7–10 bar), but greater energy per blow with an improved rate of

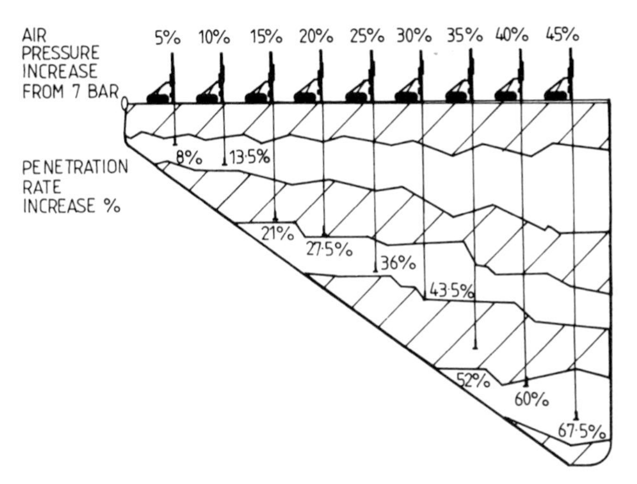

Fig. 3.27 Effect of air pressure on penetration rates.

penetration can be achieved (Fig. 3.27) by increasing the air pressure and 250 p.s.i. (18 bar) is not uncommon, especially on DTH drills. However, with the drifter method, earlier failure results from the increase in impact velocity producing considerable extra stress in the drill stem.

Bits

The penetration rate for rotary-percussive drilling is largely dependent upon rate and weight of blow, speed of rotation, length of the bit's cutting edge, and flushing efficiency. Thus there is an optimum spacing of the blow interval and rotation speed to give the most economic rate of penetration for a particular type and hardness of rock. Modern developments in tungsten carbide inserts have improved the durability of the bit, and 3 or 4 winged or cross bits (Fig. 3.28) can be used for drifter drills in most types of rock, but are particularly suited to fractured material. Other bit types (Fig. 3.29) include:

Single chisel – homogeneous rock
Double chisel – medium-hard rock
Crown bit – very hard rock
Button bit – medium to very hard rock.

The button bit (Fig. 3.30) is often selected for the DTH drill.

Button bits allow intervals between regrinding which are 4–5 times longer than for insert bits and they also tend to give better penetration rates. But insert bits are more resistant to gauge wear since they have a

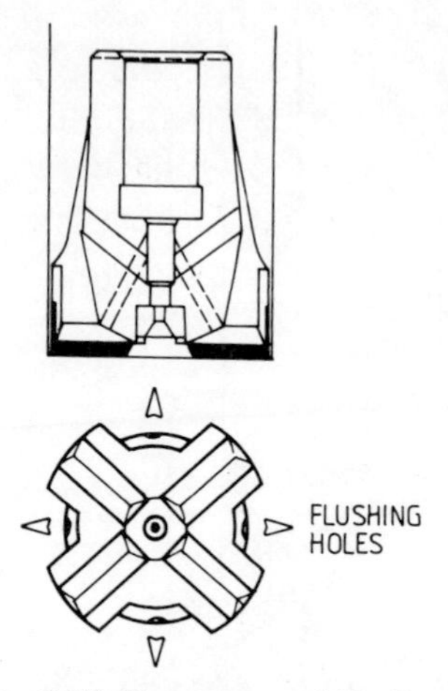

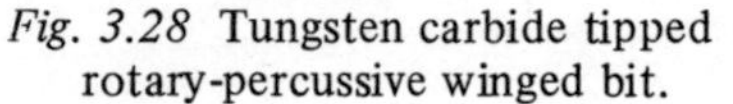

Fig. 3.28 Tungsten carbide tipped rotary-percussive winged bit.

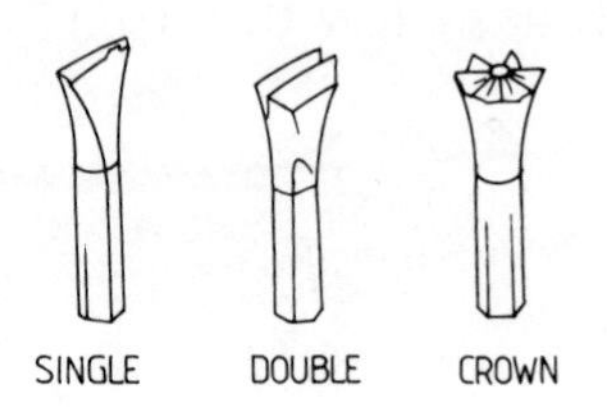

Fig. 3.29 Types of bit used in rotary-percussive drilling.

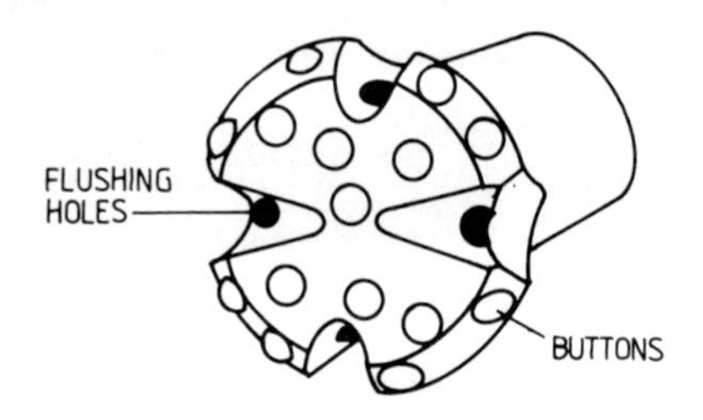

Fig. 3.30 Button bit for 'down the hole' drilling.

cemented carbide surface on the periphery. They also tend to produce straighter holes than button bits.

Drill rods for rotary and rotary-percussive drilling methods

The drill-rod is the connecting component between the drive motor on the rig and the drilling bit. Consequently, torque and feed force are transferred through the drill rod to the cutting component.

The drill-rod is hollow to allow the flushing medium to be pumped down through the drill string to the bit. It is therefore essential that the threaded joints for coupling the rods together do not allow any leakage to occur.

Rods are made in lengths to suit the particular drilling operation. For most construction operations, 3 m long rods are sufficient, but for large-scale work such as well holes etc., rods up to 7 m long are not unusual, when loading is assisted with winches.

For rotary and 'down the hole' drilling the drill stems are usually round with a large flushing hole (Fig. 3.31), but in drifter drilling, because the blow energy is transferred down the rods, a strong and rigid stem with high bending stiffnesses is required and the flushing hole is usually not larger than 10–15 mm for a 35 mm diameter rod (Fig. 3.32).

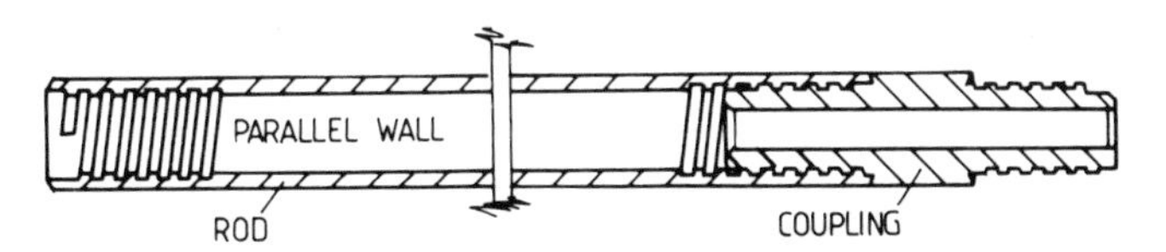

Fig. 3.31 Drilling rod for rotary drilling.

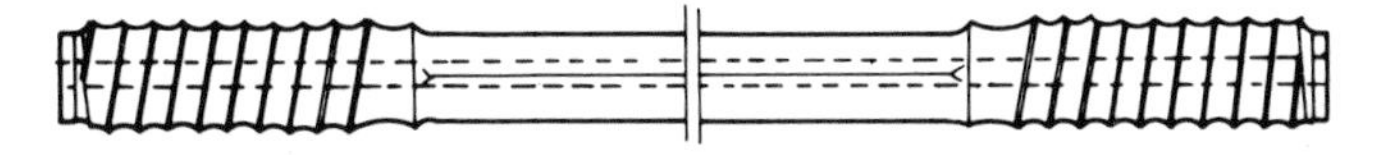

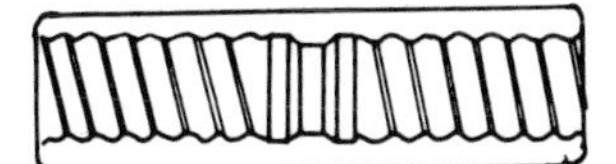

Fig. 3.32 Drilling rod for rotary-percussive drilling.

Such rods may be round or hexagonal in cross-section; a hexagonal rod helps to churn the chippings in the flushing medium and is therefore less prone to jamming.

In order to try to maintain accurate direction in deep drilling, the gap between the hole wall and the drill stem can be stabilised with guide tubes (Fig. 3.33). An accuracy of about 100 mm deviation per 33 m of drilling can normally be expected. Drifter drilling is least accurate.

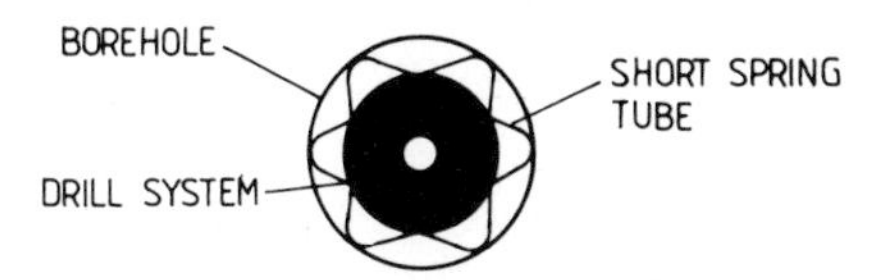

Fig. 3.33 Spacer device for drilling rods.

Flushing medium for rotary, diamond and rotary-percussive drilling methods

A flushing medium is used for two functions:

(i) To remove the cuttings from the bottom of the hole to keep the bit area free.
(ii) To cool the bit.

For percussive, rotary-percussive, or rotary methods, either water, mud or air may be chosen. Generally, water or mud is used in unstable ground, deep drilling or confined spaces and air is used where dust does not pose an environmental hazard, such as in quarrying operations. However, much of the dust and particles can be directed into a collector (Fig. 3.34) if necessary. With compressed-air drifter drills using air

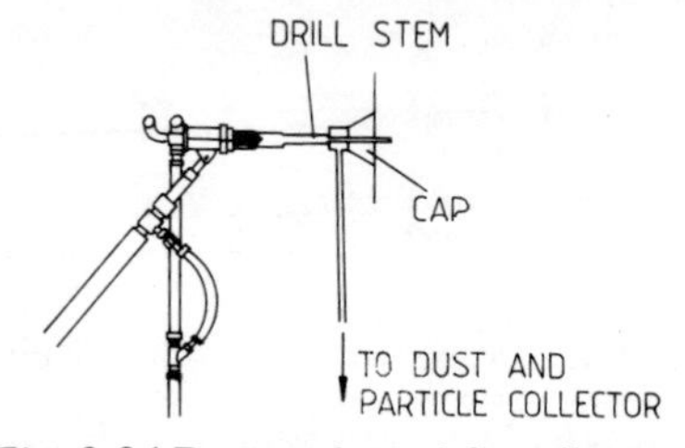

Fig. 3.34 Dust and particle collector.

flushing, the flushing pressure should not exceed that required to power the drill if the medium is fed axially through the drill, as this could cause 'back up' in the machine. Thus pressures would normally have to be less than 100 p.s.i. (7 bar) gauge. Alternatively, where greater pressures are required, for example in deep holes, a separate swivel can be attached to the drill string and pressures of up to 200 lb/in^2 gauge or more (14 bar) can then be achieved.

If cleaning is to be effective, then for water or mud the flushing speed should be between 0.4 and 1.0 m/s (0.3–0.6 m/s for diamond drilling) depending upon the size and density of the chippings. For air the ideal rate should be 15–30 m/s in order to lift and carry the particles. By reference to Fig. 3.35, the quantity of flushing medium passing through the annulus area is

$$Q = \frac{\pi}{4}(D^2 - d^2) \times U \times 60$$

where Q = m^3/min
U = m/s
d = drill pipe outside diameter in m
D = borehole diameter in m

Thus for a 100 mm diameter hole, produced by a drifter drill using air flushing with 35 mm diameter rod (couplers), when $U = 15$ m/s

$$Q = \frac{\pi}{4}(0.1^2 - 0.035^2) \times 15 \times 60$$

$$= 6 \text{ m}^3/\text{min } (214 \text{ ft}^3/\text{min})$$

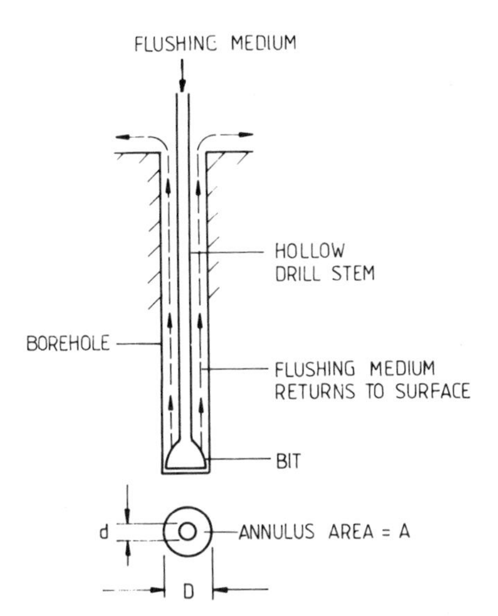

Fig. 3.35 **Drill-hole flushing action.**

If air is adopted as the flushing medium, the gauge air pressure is usually 7 bar. For a 13 mm diameter hole in the drilling rods, only about 2.25 m^3/min of air at this pressure will reach the bit when passed down a 30 m stem. Therefore the corresponding annulus velocity is

$$U = \frac{2.25}{\frac{\pi}{4}(0.1^2 - 0.035^2) \times 60}$$

$$= 5.4 \text{ m/s}$$

As a consequence, the chippings must be broken down into very small pieces before they can be carried away, thereby slowing down the drilling rate.

Flushing is less of a problem with the DTH drill, because the drill rods are much larger in diameter, thus reducing the size of the annulus. Furthermore, the hole in the rods which serves to supply air to drive the piston can be greater and subsequently used for flushing. Higher rates of penetration per kW of power supply are therefore possible with DTH than with drifter drilling. For similar reasons flushing is also generally not a significant problem with the rotary drilling method.

Drilling muds and foams

The consistency of drilling mud depends upon the hole stability, the density and size of cuttings and the rate of penetration, but it is usually of a creamy to custard consistency. Natural mud is generally a bentonite clay which exhibits thixotropic characteristics, but it can also be made from biodegradable organic polymers. To reduce the quantity of flushing medium required, the mud is passed through tanks (Fig. 3.36)

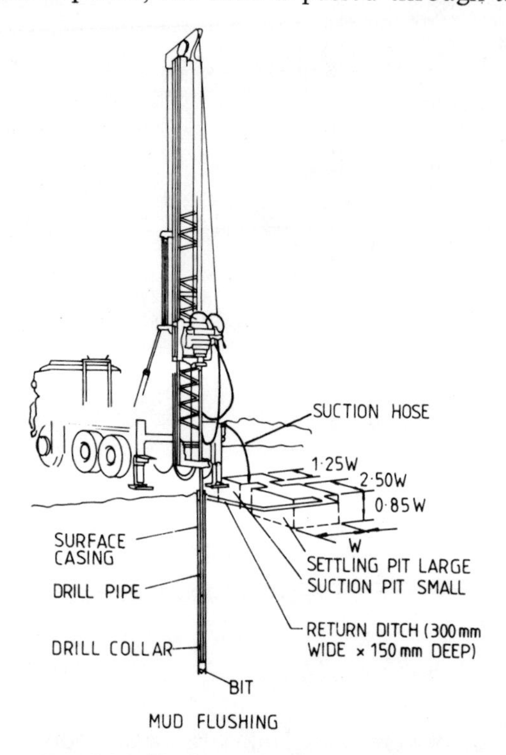

Fig. 3.36 **Flushing medium settling tanks.**

– a suction pit and a settling pit. Mud is pumped from the suction pit down the hole and is returned to the settling pit, where the cuttings are deposited. Mud then flows back to the suction pit.

The pits are commonly dug in the ground and the volume of the pits should be at least three times the anticipated drill hole volume. The mud density and viscosity should be continuously tested during use.

Foam may be used as an alternative to mud since it usually has greater carrying capacity than water-based muds and produces a lower hydrostatic head.

Foam is made by forming an emulsion between water and a foaming agent, which can be natural liquid soap or polymer-based. The consistency used will depend on the amount of support required and density of cuttings to be lifted, but it is commonly of 'shaving foam' consistency. The velocity of the foam is between 0.23 and 0.75 m/s. Normal foam is usually a 1% volume mix.

Other forms of rotary-percussive drilling equipment

Drills and drilling rigs for tunnelling applications

Drilling holes in the face or sides of a tunnel requires a more versatile and manoeuvrable rig than that used primarily for vertical drilling. Furthermore, to achieve high levels of production it is often necessary to produce several boreholes at the same time. Multi-head rigs have been developed to suit this need, supported either by rails (Fig. 3.37) or on rubber-tyred wheels (Fig. 3.38). This method, using a rotary-percussive action, is usually referred to as drifter drilling, as opposed to bench drilling in quarrying.

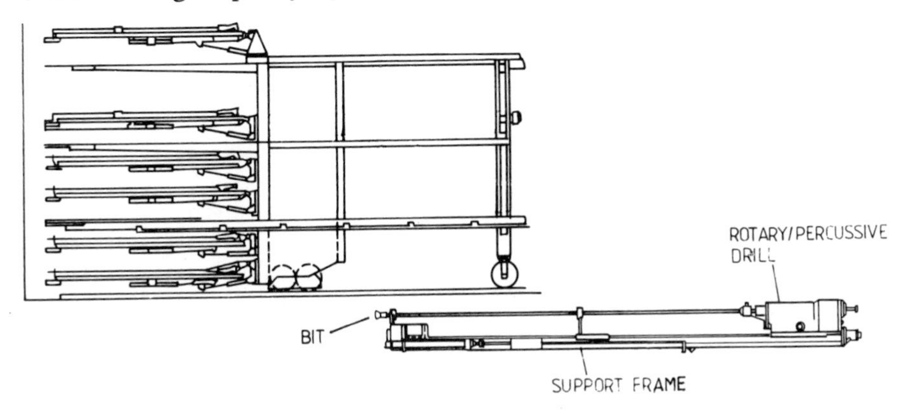

Fig. 3.37 Rail-mounted multi-boom drifter drilling rigs.

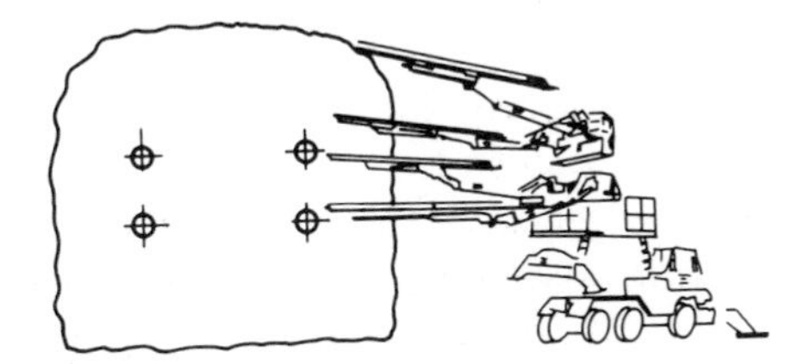

Fig. 3.38 Wheel-mounted multi-boom drifter drilling rigs.

Hand drills *(Fig. 3.39)*

The hand-held drill is very similar in appearance to the concrete breaker, except for the rotary action; it weighs 10–20 kg, and is suitable for drilling shallow and small diameter holes, up to about 30 mm diameter,

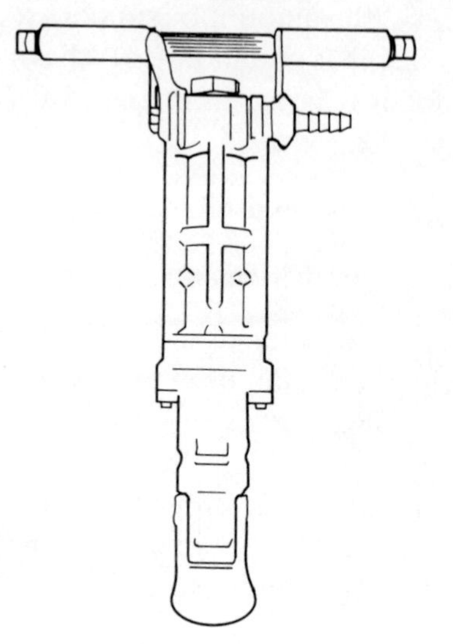

Fig. 3.39 Hand held percussive drill.

to depths not exceeding approx. 8 m.

It operates on the combined rotary-percussive principle at approximately 145 r.p.m./2000 blow p.m. and requires about 1–2 m^3/m free air supply (including air for flushing). However, water may also be used as a flushing medium.

All machines are fitted with a detachable muffler, but the operator should also wear ear defenders.

Jack and feedleg drills

These are similar to hand drills, except that the drill uses the thrusting action of a feed leg, hydraulically controlled to provide feed force. Reaction to the thrust is either against an opposite wall or from an operator's board (Fig. 3.40). Higher penetration rates can be achieved because of the extra control offered by the feed leg, which can also be

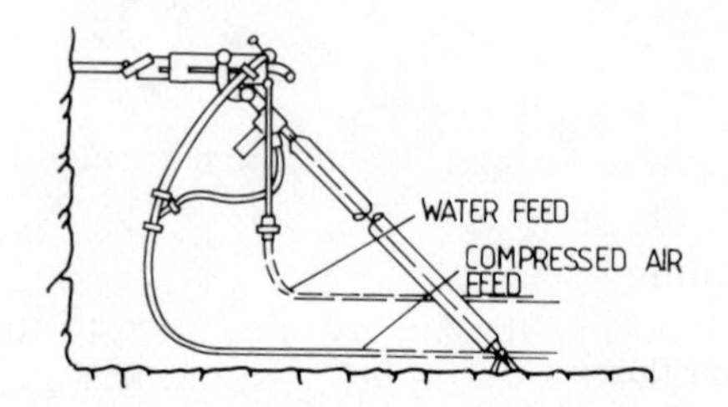

Fig. 3.40 Jack and feedleg rotary-percussive drill.

reversed to pull out the drill rod in case of jamming. The equipment is fairly light (90 kg) and is fitted with handles which make it easily transportable.

Jammed bits

When using rotary-percussive methods, there is a tendency that the bit will get stuck, especially if it is of the older type, without tungsten carbide inserts. Such bits can only penetrate a metre or so, depending upon the rock hardness, before requiring replacement. To overcome the jamming problem it is advisable to taper such holes in steps by changing the bit size, as shown in Fig. 3.41. Practice has demonstrated that the diameter of the hole should be reduced by about 3 mm for each metre of borehole length.

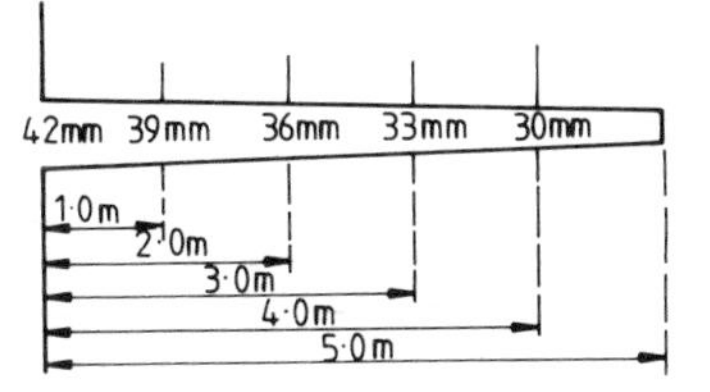

Fig. 3.41 Taper borehole.

Rotary-percussive drills data

(i) Drifter drills

Hole diameter (mm)	*Drill weight* (kg)	*Air consumption* (m^3/min at 7 bar)	*Feed type*
up to 30 mm	10	1.5	Hand
up to 40 mm	20	3.0	Hand
up to 40 mm	40	4.5	Feed leg
up to 60 mm	90	7.5	Chain feed
up to 90 mm	150	12.0	Chain feed
up to 130 mm	250	18.0	Chain feed

(ii) DTH drills

up to 100 mm	40	8*	Chain feed
up to 150 mm	80	13	Chain feed

*Excluding air flushing requirement.

Production data for rotary-percussive drilling

Table 3.7 Drilling performance – 100 mm diameter drifter drill

Rock type	Bit loading (N/mm bit dia)	Penetration rate (m/h)	Bit life (m. of hole)
Soft	100-200	30-50	300-500
Hard	100-200	10-20	100-350
Very hard	100-200	5-10	50-150

The 100 mm diameter DTH drill, with about $\frac{1}{4}$ the power of the drifter drill, achieves approximately $\frac{1}{3}$ the above rates of penetration, when drilling holes down to 30 m or so.

Percussive drilling

Rope-operated rig *(Fig. 3.42)*

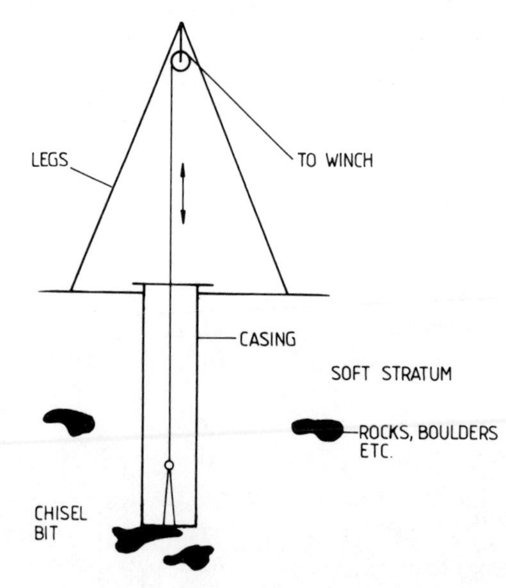

Fig. 3.42 Rope operated percussive drilling.

When rope-operated rigs are being used in soil investigation, pile-driving, etc., it is sometimes more expedient to use a chisel suspended from a rope and operated by a winch to break through a thin seam or intrusion of soft rock.

Hand tools *(Fig. 3.43)*

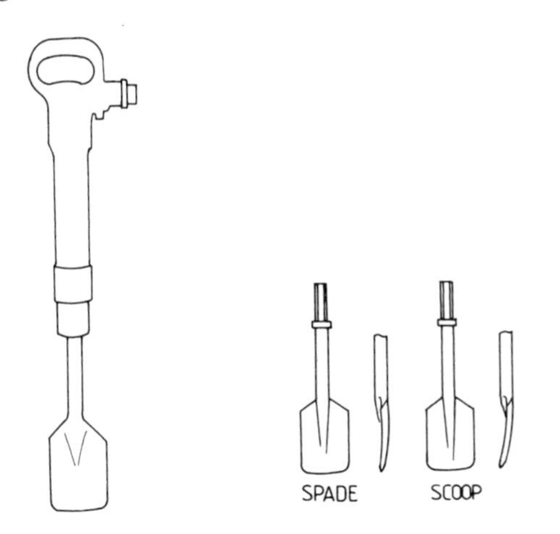

Fig. 3.43 Air driven hand tools.

Small powered tools are required for concrete breaking, roadworks, picks, spades, etc. These are usually light, 30–40 kg, and use up to 1-2 m^3/min air supply. However, these small tools are increasingly using hydraulic powering. The method relies purely on percussion; rotary action is not included.

References

Aleman, V.P. (1981) 'A strata index for boom type road headers'. *Tunnels and Tunnelling,* **13** (2), (March).

American Petroleum Institute (STD 13A, 1969 – RP 13B, 1972) *Drilling fluid materials.*

Antonor, L.N. (1980) 'Effective technology for drilling and blasting work'. *Razvedka i Okhrana*, no. **1**, (January).

Atlas Copco Manual, 2nd Edition (1975). Stockholm, Sweden.

Brennstein, E. (1979) 'Studies of the relations between main parameters during percussive drilling with hydraulic percussive drills'. *Berg- und Huettenmaennische, Mondatshefle*, **124** (10), (October).

Broomwade Ltd. Trade literature. High Wycombe, Bucks.

Brunsing, T.P. (1980) 'The effect of tensile pre-stress on the efficiency of drag bit cutting in hard rock'. Ph.D. thesis, University of California, Berkeley, U.S.A.

Bullock, R.L. (1976) 'Survey of hydraulic drilling performance'. *Tunnels and Tunnelling,* **8** (6), (September).

Chicago Pneumatic Company. Trade literature. Enid, Oklahoma, U.S.A.

Christensen Diamond Products Co. Trade literature. Ashford, Middlesex.

CIRIA Report 79. 'Tunnelling–improved contract practices' (1980).

Clemmor, R.J. (1965) 'The design of percussive drilling bits'. *Mining and Minerals Engineering* (February and March issues).

Consolidated Pneumatic (Holman) Co. Ltd. Trade literature. Fraserburgh, Aberdeenshire.

Craelius Ltd. Trade literature. Daventry, Northants.

Crimmins, R. (1972) *Construction Rock Work Guide*. Wiley & Son.

Dickenson, E.H. (1960) *Rock Drill Data*. Ingersol–Rand Co., New York.

Ellerton, E.J. (1982) 'Equipment and methods for rock drilling'. B.Sc. Project Report, Department of Civil Engineering, Loughborough University of Technology.

England, R.A. (1979) 'Circulation fluids for water well drilling'. *Ground Engineering*, (April).

English Drilling Equipment Ltd. Trade literature. Huddersfield.

Gardner–Denver Company. Trade literature. Quincy, Illinois, U.S.A.

Hands–England Drilling Ltd. Trade literature. Letchworth, Herts.

Hughes Tool Company. Trade literature. Houston, Texas, U.S.A.

Ingersoll–Rand Company. Trade literature. Hosksville and Phillipsburg, N.J., U.S.A.

Longyear Company. Trade literature. Delaware, U.S.A.

MacFeat–Smith, I. (1979) 'Quantification of the cutting abilities of road header tunnelling machines'. *Tunnels and Tunnelling,* **11** (Dec.).

McGregory, M.K. (1968). *The drilling of rock*. C.R. Brooks Ltd., A. MacLaren Co.

Marx, C. (1970) *Diamond bits and their use in shallow holes*. Christensen Diamond Products Co., Mining Division, Germany.

Maurer, W.C. (1977) 'Advanced excavation methods'. *Underground Space,* **2** (2), (December).

National Lead Co. (Baroid Division) (1964) *Drilling mud data book*. Houston, Texas, U.S.A.

Nelmark, J.D. (1980) 'Large diameter blast hole drills'. *Mining Congress Journal,* **66**, (August).

Persen, L.N. (1975) *Rock dynamics and Geophysical exploration*. Elsevier.

Persson, P.A. and Schmidt, R.C. (1977) *Mechanical boring and drill and blast tunnelling*. Stockholm.

Pohle, G. (1976) *Maschinen und Verfahren des Banbetriebes*. Aachen Technische Hochschule, Germany.

Rehbinder, G. (1980) 'Theory of cutting rock with a water jet'. *Rock Mechanics,* **12** (3–4), (March).

Rock Drilling Manual (1977) Theory and Technique. Sandvik, A.B., Sweden.

Tamrock Ltd. Trade literature. Tampere, Finland.

CHAPTER FOUR

Rock blasting

Introduction

The use of explosives to break up rock and hard soils is required in tunnelling, quarrying and for excavation works where mechanical plant cannot perform the task economically. Explosives are also used in the demolition of many types of structure, such as bridges, foundations, high rise buildings, chimneys, cooling towers, etc.

Explosion in soil and rock

The effect of an explosion is to convert a chemical substance into a gas which then produces enormous pressure. The process takes place rapidly in 2000–6000 m/s shattering the rock adjacent to the explosive and exposing the surrounding area to stress.

The nature of the wave formations presents the following phenomena:

On ignition of a spherical charge placed in a solid medium such as rock, a shock wave is propagated, causing crushing and possibly liquefaction of the adjacent rock. Radial cracks are also formed which fade out with increasing distance from the wave front, as the energy is dispersed to a greater volume of the medium. In most applications of explosives in construction work, a free face is usually present and a more complex pattern of wave formation is produced, as shown in Fig. 4.1. On meeting the free surface (Fig. 4.1(a)) the initial shock wave travelling out from the centre of the explosion is reversed in direction and moves away from the epicentre O^1 (Fig. 4.1(b)). On reaching the explosion cavity, the wave front is again reflected, which, together with the force of the expanding gases, raises the medium into a cupola (Fig. 4.1(c)). If the charge is sufficiently powerful, then material is forced outwards to produce a pattern of shattering and cracking of the rock as illustrated in Fig. 4.2.

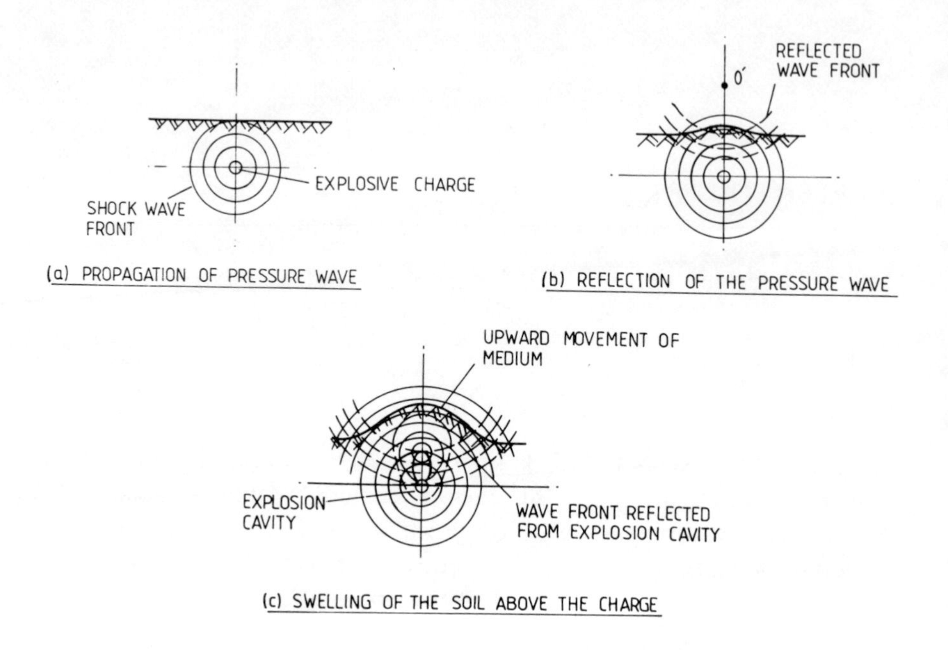

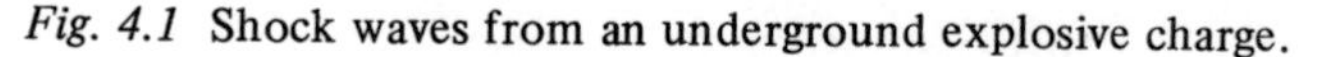

Fig. 4.1 Shock waves from an underground explosive charge.

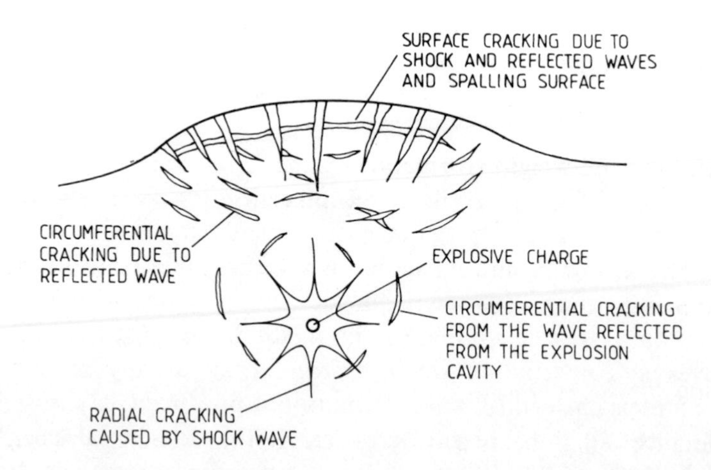

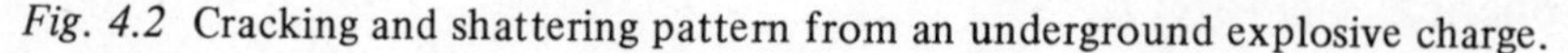

Fig. 4.2 Cracking and shattering pattern from an underground explosive charge.

Blasting theory

An explosive charge placed at a depth B (called the burden) below a free surface is assumed to throw out materials to form a conically shaped crater (Fig. 4.3) with 45° side slopes. The energy required to impart the initial velocities will be proportional to the mass of the body

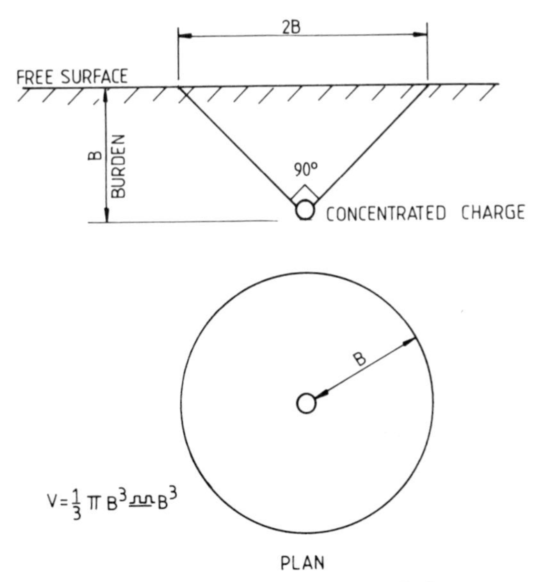

Fig. 4.3 Effects of a concentrated charge.

of material in the cone, i.e. B^3. If gravitational forces are present, the energy required to lift the mass to surface level is proportional to the product of the mass and the distance moved (B) (i.e.) B^4. The energy required to overcome the frictional forces between the surface of the crater and surrounding medium is proportional to the crater surface area (i.e. B^2).

Thus, the total weight of charge required is given by

$$Q = c_2B^2 + c_3B^3 + c_4B^4 \tag{4.1}$$

(*Note*: The use of c_0 and c_1 has been avoided in order to maintain the same nomenclature as other writers.)

Benching

Equation (4.1) has been shown by Langefors to be appropriate for a spherical charge (Fig. 4.3) and for a section of cylindrical charge of length B, where B is much smaller than the full length of charge (h) (Fig. 4.4). Thus to tear the burden in the case of benching, the charge in the drill hole may be considered to consist of two partial charges:

(i) a concentrated spherical charge at the bottom
(ii) a cylindrical charge for the rest of the column (neglecting end effects).

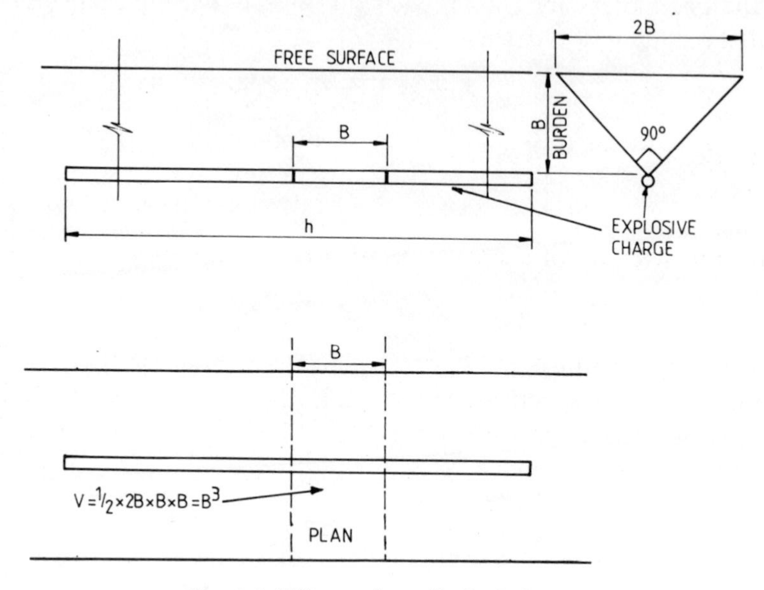

Fig. 4.4 Effects of a cylindical charge.

(i) Bottom charge (Fig. 4.5)

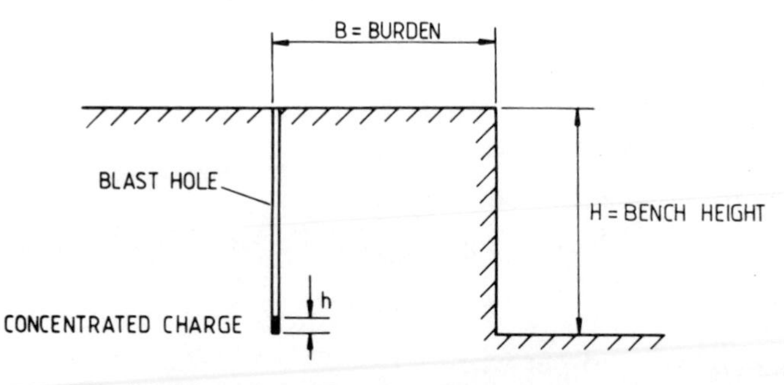

Fig. 4.5 Fixed base blasting with a concentrated charge.

The concentrated bottom charge Q_1 given by eqn. (4.1) is

$$Q_1 = c_2B^2 + c_3B^3 + c_4B^4 \qquad (4.2)$$

where Q_1 is in kg of a given explosive
B is the burden (depth of charge) in metres
c_2, c_3, c_4 are empirical coefficients which depend on the bench height (H), burden (B) and length of explosive charge (h)

whereby

$$c_{\text{i}} = k_{\text{i}}\left(\frac{H}{B}, \frac{h}{B}\right)$$

If the parameters are chosen such that $H = B$ and the charge is concentrated so that $h/B \triangleq 0$, then

$$c_i = c_i(1,0) \text{ and so } c_2(1,0), c_3(1,0), c_4(1,0)$$

These may be represented by new constants

$$a_2, a_3 \text{ and } a_4$$

Thus the quantity of charge Q_1 needed to tear the burden is given by

$$Q_1 = a_2B^2 + a_3B^3 + a_4B^4 \tag{4.3}$$

Values for a_2 and a_3 can be determined from test blasts with B at say 0.5 m and 1 m respectively and the simultaneous equations solved. This task is made simpler by ignoring a_4 which produces less than 1% of Q_1 for a bench arrangement.

(ii) Column charge (Fig. 4.6)

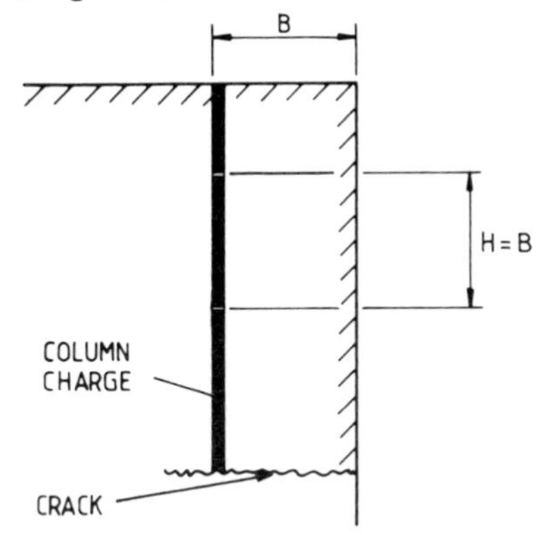

Fig. 4.6 Free base blasting with a cylindrical charge.

For a high bench, a section of length $H=B$ of the column can be considered to be cylindrically charged (see Fig. 4.5). Thus the weight of charge (Q_2) required to tear the burden is given by

$$Q_2 = b_2B^2 + b_3B^3 + b_4B^4 \tag{4.4}$$

where b_i is the special case of k_i (for a cylindrical charge).

The quantity of charge required per metre (q) for a section of bench of height equal to the burden (B) is therefore

$$\frac{Q_2}{B} = q = b_2B + b_3B^2 + b_4B^3 \tag{4.5}$$

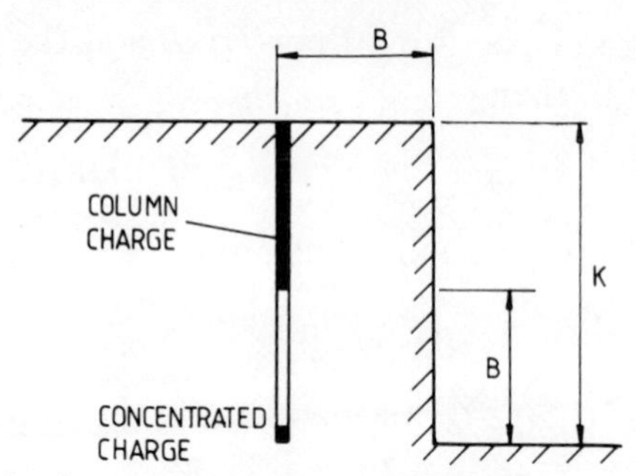

Fig. 4.7 Combined concentrated and cylindrical charge for bench blasting

The total charge for a bench of height k (Fig. 4.7) is therefore a summation of the concentrated charge at the base Q_1, and the column charge (q) which is assumed to be proportional to $(k - B)$ (neglecting end effects). Thus

$$Q_T = Q_1 + q(k-B) \tag{4.6}$$
$$= a_2B^2 + a_3B^3 + a_4B^4 + (b_2B + b_3B^2 + b_4B^3)(k-B) \tag{4.7}$$

Langefors indicates that tests have demonstrated that for benching, $b_2 = 0.4a_2$ and $b_3 = 0.4a_3$ irrespective of rock type, thus rearranging (4.7),

$$Q_T = 0.4a_2\left(\frac{k}{B} + 1.5\right)B^2 + 0.4a_3\left(\frac{k}{B} + 1.5\right)B^3 + \left(a_4 + b_4\left(\frac{k}{B} - 1\right)\right)B^4 \tag{4.8}$$

where a_2, a_3 and a_4 are determined for a bottom charge in case (i).

In practice the bottom charge cannot be concentrated and has to be distributed in the drill hole uniformly from the bottom upwards and therefore the full effect of a concentrated bottom charge cannot be realised. Tests by Langefors indicate that in this type of loading the charge effect at the base is only about 60% of the theoretical value with a concentrated charge. However, it was shown that the breaking power could be increased to almost 90% of the theoretical value by drilling the hole to a depth $0.3B$ below the base and loading the bottom charge to a height of $1.3B$. The bottom charge distributed in this manner was equivalent to Q_T in equation (4.8) for a bench height of $k = 2B$. Thus the required uniformly distributed bottom charge (Q_b) can be calculated from (4.8)

for $k = 2B$, as

$$Q_b = 1.4a_2B^2 + 1.4a_3B^3 + a_4B^4 + b_4B^4 \tag{4.9}$$

When the bench height (k) exceeds $2B$, the additional charge required can be calculated as a column charge from equation (4.5). Thus the column charge Q_c is

$$Q_c = q(k - 2B) = (b_2 B + b_3 B^2 + b_4 B^3)(k - 2B)$$

and since $b_2 = 0.4a_2, b_3 = 0.4a_3$

$$Q_c = 0.4\left(\frac{k}{B} - 2\right)(a_2 B^2 + a_3 B^3) + b_4(k - 2B)B^3 \qquad (4.10)$$

adding equations (4.9) and (4.10),

$$Q_T = Q_b + Q_c = 0.4a_2\left(\frac{k}{B} + 1.5\right)B^2 + 0.4a_3\left(\frac{k}{B} + 1.5\right)B^3 + \left(a_4 + b_4\left(\frac{k}{B} - 1\right)\right)B^4 \qquad (4.11)$$

where Q_T = Total charge

which also corresponds with equation (4.8).

Note: a_2 and a_3 are determined experimentally as described in case (i) and b_4 is very small. (For usual cases with a standard dynamite these are approximately 0.07 kg/m^2, 0.4 kg/m^3, 0.004 kg/m^4 respectively.) The distribution of the complete charging is shown in Fig. 4.8.

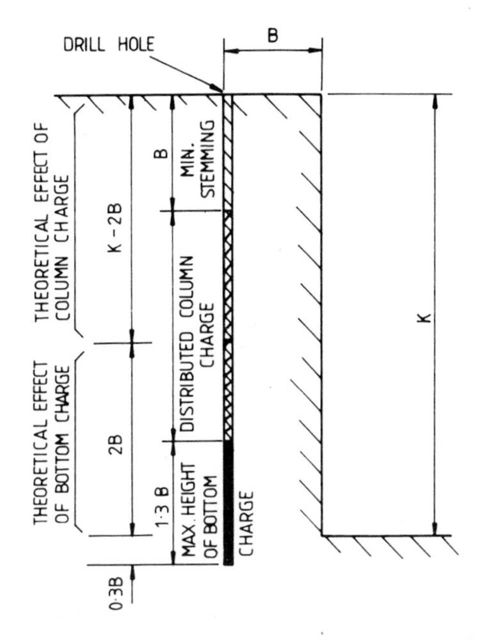

Fig. 4.8 Distribution and effect of explosive for bench blasting

Effect of charge strength

Clearly, for the case when $k \leqslant 2B$, the bottom charge would cause excessive throw and the burden must therefore be reduced (and the charge recalculated). Alternatively the use of an explosive allowing a more concentrated charge at the bottom would be more favourable, as described under sections (i) and (ii). For these reasons, the maximum height of the bench is generally limited to about 30 m in quartz as a safety precaution.

The use of a stronger explosive can be incorporated into equations (4.9), (4.10) and (4.11) by the introduction of a factor $\frac{1}{w}$ where

$$\frac{1}{w} = \frac{\text{specific energy (unit weight strength) of the standard explosive}}{\text{specific energy (unit weight strength) of the explosive used}}$$

therefore

$$Q_b' = \frac{1}{w} Q_b \tag{4.12}$$

$$Q_c' = \frac{1}{w} Q_c \tag{4.13}$$

$$Q_T' = \frac{1}{w} Q_T \tag{4.14}$$

Effect of hole spacing

Where a row of drill holes is involved, the volume of material available for removal by each blast is less than for a single drill hole. For example in Fig. 4.9, with $B = S$ there is a degree of overlapping and tests have demonstrated that the charge in each drill hole may be reduced by about 20%.

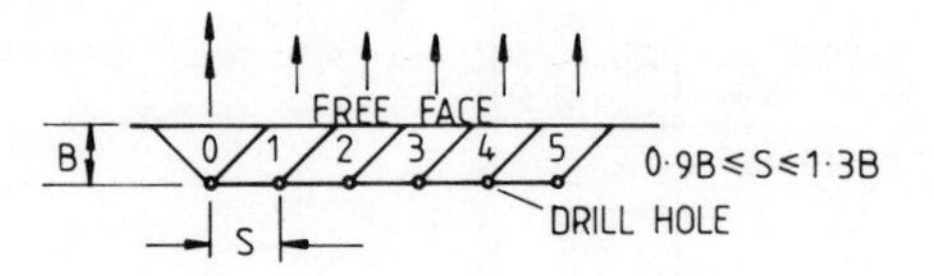

Fig. 4.9 Single row short delay blasting

Thus $$Q_T'' = \frac{0.8}{w} Q_T \text{ when } B = S.$$

For other spacings Q_T has been found to be proportional to the ratio $\frac{S}{B}$.

Thus
$$Q_T'' = \frac{0.8}{w} \cdot \frac{S}{B} \times Q_T \quad (4.15)$$

Generally the ratio is held within the bounds $0.9 \leqslant \frac{S}{B} \leqslant 1.3$. At the lower limit fragmentation is poor, whilst at the upper limit uneven tearing may result.

Effect of slope and fixity

The charge required for a vertical bench with a free base (Fig. 4.6) is about 75% of that for a fixed base (Fig. 4.5) and is incorporated into the charge formula by a degree of fixity (f) factor. Also, less energy is required to tear the burden of a sloping face, because of the larger angle at the base.

Langefors suggests the following values for f

	Vertical bench	*Sloping bench* 3:1	*Sloping bench* 2:1
Free base	0.75	0.75	0.75
Fixed base	1.00	0.9	0.85

Thus
$$Q_T''' = f \times \frac{0.8}{w} \times \frac{S}{B} \times Q_T \quad (4.16)$$

Effect of hole diameter

The hole diameter has little effect on the degree of loosening as it is the total quantity of charge, particularly the bottom charge, which determines the breakage.

Example

A drill hole of diameter d(mm), with the bottom charge extending $1.3B$ from the base, contains explosive of unit weight $P\,\text{kg/m}^3$, thus

$$\frac{P}{10^6} \times \frac{\pi d^2}{4} \times 1.3B = Q_b''' =$$
$$f \times \frac{0.8}{w} \times \frac{S}{B} (1.4a_2 B^2 + 1.4a_3 B^3) + (a_4 + b_4) B^4 \quad (4.17)$$

When $f = 1$, $\frac{S}{B} = 1$, $w = 1$, $P = 1000$, $a_2 = 0.07$, $a_3 = 0.4$ and terms with B^4 are ignored

$$\frac{d^2}{0.846 \times 36^2} = B^2\left(\frac{0.07}{B} + 0.4\right)$$

or $$B^2 \triangleq \frac{1}{0.86} \times \left(\frac{d}{36}\right)^2, \text{ since } \frac{0.07}{B} \text{ is very small,}$$

$$B \triangleq \frac{1}{\sqrt{0.86}} \times \frac{d}{36} \triangleq 0.03d$$

where B is in m and d in mm e.g. when $d = 50$ mm $\underline{B = 1.5\text{ m}}$.

Throw and scattering (Fig. 4.10)

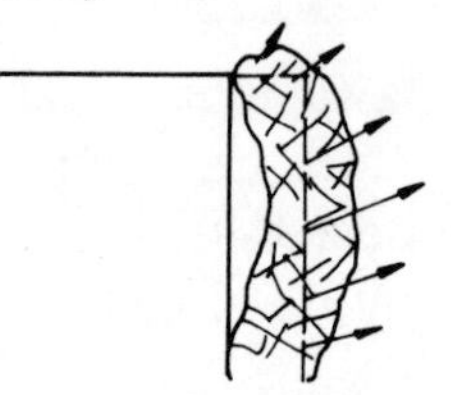

Fig. 4.10 Throw of debris

An increase in the charge causes increased cracking and shattering of the rock. In addition material will be thrown forward, including fragmentation and scattering. Calculations can be made to determine the extent of the effects and the reader is referred to Langefors and Kihlstroem for a detailed discussion.

Example to determine total quantity of charge (Fig. 4.11)

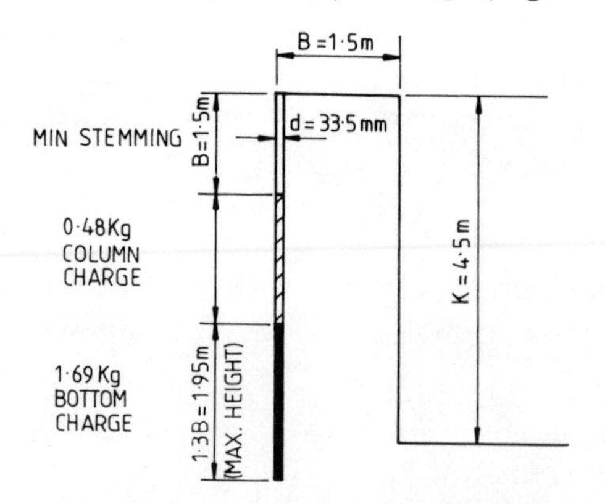

Fig. 4.11 Example of distribution of explosive charges

Assume $P = 1000\text{ kg/m}^3$, $\frac{S}{B} = 1$, $w = 1$, $a_2 = 0.07\text{ kg/m}^2$, $a_3 = 0.4\text{ kg/m}^3$ and terms with B^4 can be ignored. $B = 1.5$ m and $k = 4.5$ m.

(i) Bottom charge
From equation (4.9)

$$Q_b''' = f \times \frac{0.8}{w} \times \frac{S}{B} (1.4a_2B^2 + 1.4a_3B^3)$$

$$= 0.8\,(0.225 + 1.89) = \underline{1.69\text{ kg}}.$$

(ii) Diameter of drill hole

$$\frac{P}{10^6} \times \frac{\pi d^2}{4} \times l = 1.69 \quad \text{if } l = 1.3B = 1.3 \times 1.5 = 1.95\text{ m}$$

$$d^2 = \frac{1.69}{1.95} \times \frac{4}{\pi} \times 1000 = 1104$$

$$d = \sqrt{1104} = \underline{33.5\text{ mm drill hole}}$$

(iii) Column charge

$$Q_c''' = f \cdot \frac{0.8}{w} \cdot \frac{S}{B} \left[0.4\left(\frac{k}{B} - 2\right)(a_2B^2 + a_3B^3) + b_4(k - 2B)B^3\right]$$

$$= 0.8 \times 0.4 \times (0.16 + 1.35) = \underline{0.48\text{ kg}}$$

Blasting patterns

When the charge requirements have been calculated, the blast hole pattern can be designed. The method of firing is performed in one of two ways:

(i) by simultaneous shots
(ii) by short delay blasting

Simultaneous method

In this method a number of holes are fired simultaneously and it is therefore ideal for breaking out a long bench of rock. The technique requires less accuracy than delay shots, and a slight variation in hole spacing and direction of the line of the drill hole can be accepted. Excessive ground vibration, scatter and overbreak may occur, but when quarrying for rough stone these problems may not be important.

Short delay method

(i) Single-row short delay blasting
Practice has demonstrated that by delaying the shots in sequence, the

result is improved fragmentation, reduced throw and a slightly lower consumption of explosive. Langefors and Kihlstroem suggest that the delay interval t can be related to the burden (B) by the formula

$$t = cB$$

where t is in m/s
B is in m
c is a constant $\triangleq$ 3 micro seconds/metre

Several different orders of sequence have been developed depending upon the conditions, but for a homogeneous material, with accurately timed and spaced drill holes, the sequence in Fig. 4.9 is often used.

(ii) Multiple-row short delay blasting
The principle is similar to instantaneous single-row blasting, but with several rows shot in a delay sequence (as shown in Fig. 4.12). The effect

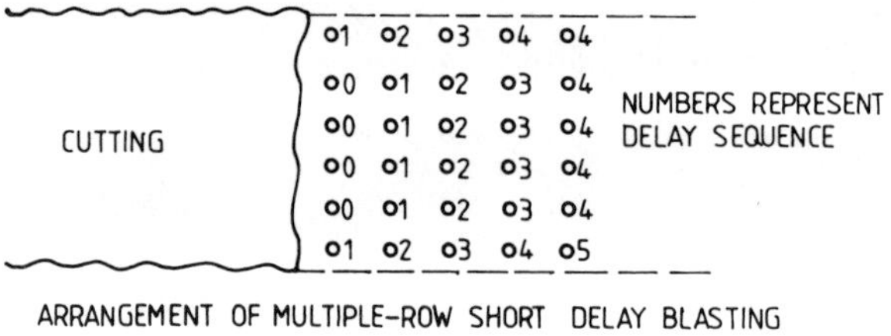

CUTTING

o1 o2 o3 o4 o4
o0 o1 o2 o3 o4
o0 o1 o2 o3 o4
o0 o1 o2 o3 o4
o0 o1 o2 o3 o4
o1 o2 o3 o4 o5

NUMBERS REPRESENT DELAY SEQUENCE

ARRANGEMENT OF MULTIPLE-ROW SHORT DELAY BLASTING

Fig. 4.12 Arrangement of multiple-row short delay blasting

is to produce a more intense fragmentation, behind the first row, and significantly reduce flying rock. This method is often used for foundation work and also in quarrying, where large quantities of rock are required to keep the work gangs and all equipment fully utilised.

In blasting large excavations, a preliminary area must be loosened first to produce a free face for subsequent benches. This first, or sinking, cut should be heavily charged as only a single free face is available and the shots should be delayed in sequence as shown in Fig. 4.13. The area may be subsequently expanded, section by section, using either single row delays, i.e. removing the material progressively (Fig. 4.9) or with the multiple-row technique (Fig. 4.14). Material from the sinking cut may either be excavated or left in place. In the multiple-row technique, when the subsequent firings take place the ground will have been loosened sufficiently to allow tearing of the burden.

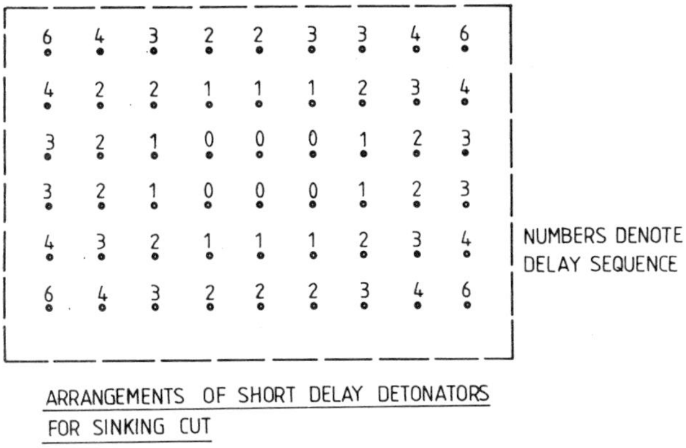

6 4 3 2 2 3 3 4 6
4 2 2 1 1 1 2 3 4
3 2 1 0 0 0 1 2 3
3 2 1 0 0 0 1 2 3
4 3 2 1 1 1 2 3 4
6 4 3 2 2 2 3 4 6

NUMBERS DENOTE DELAY SEQUENCE

ARRANGEMENTS OF SHORT DELAY DETONATORS FOR SINKING CUT

Fig. 4.13 Short delay blasting for a sinking cut

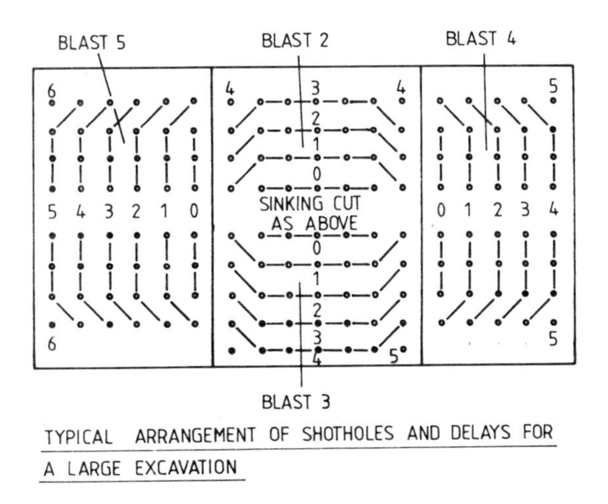

Fig. 4.14 Arrangement of delays for a large excavation

Prevention of overbreak

The above blasting methods unfortunately produce overbreak and irregularities. A smoother wall for cuttings and tunnels can be obtained by pre-split blasting. This technique consists of lightly charged and well stemmed shots placed around the perimeter of the excavation at close centres (500 mm). The drill hole is usually of small diameter (50 mm max.) and must be accurately drilled, to give even effect.

The pre-splitting charge is exploded, followed either instantaneously or with a slight delay by the main charges.

Tunnel blasting

The principles of calculating the required charges in tunnel blasting are similar to those described for benching. The charges are arranged in a predetermined pattern and released in a planned sequence as in short delay blasting.

The first holes in the sequence produce the loosening, thus providing the free face for subsequent opening up of the heading. Two principal methods have been developed, namely wedge cuts and parallel hole cuts. The latter is easier to carry out, but is not well suited to soft rock.

Wedge cuts (*Fig. 4.15*)

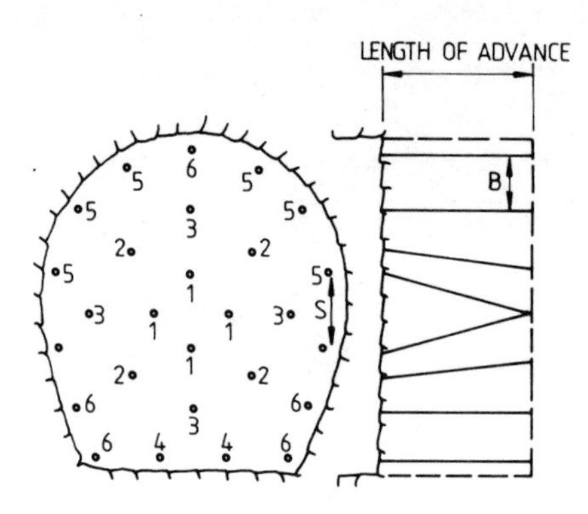

Fig. 4.15 Arrangement of wedge/fan cut for a tunnel heading

The holes are arranged regularly to facilitate drilling and are set with short delays usually progressing outward to form a fan, wedge or conical section at the centre of the face, which is thereby released first to provide further free faces for the subsequent rows of charges.

Parallel cuts (*Fig. 4.16*)

In recent years successful results have been obtained by blasting towards several closely spaced, uncharged, empty holes, or alternatively towards a single, large (100 - 150 mm diameter) uncharged hole. The effect is to produce a progressive enlargement along the full advance. This method has the advantage that the holes can be drilled parallel to each other with a considerable saving in time and expertise. However, in soft rock or where the advance becomes excessive, the rock tends to jam and the opening-up effect is hindered.

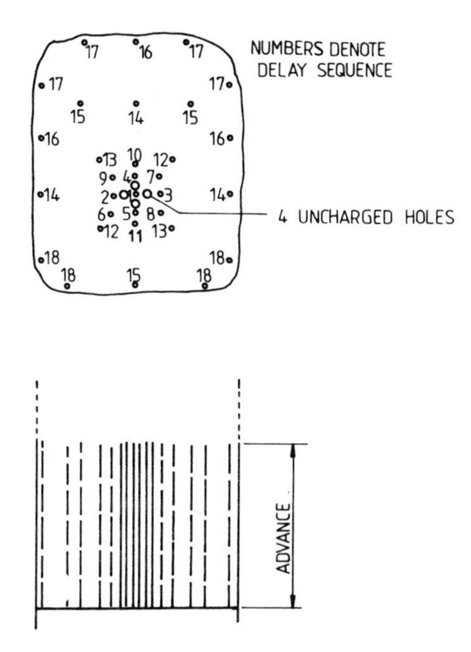

Fig. 4.16 Arrangement of a parallel cut for a tunnel heading

Length of advance

The length of drive to be selected depends upon the span of rock which can remain unsupported during the time taken to install the supports. However, by using a modern method of temporary support such as guniting, rock bolting, ribbing, (described in Chapter 10) full face advance within the range 2 – 7 m is commonly obtained. As a rule of thumb, however, the length of advance is usually not greater than the face width.

In very poor strength rock, where the propping could not be fixed over the full face in the time available before a collapse occurs, it may be necessary to divide the face up into sections, whereby the roof would be propped and the sides removed in stages.

Generally full face tunnelling produces the lowest costs, but where the face cross-sectional area exceeds about 50 m², hole drilling with the modern multi-headed jumbo drills and subsequent blasting are best achieved from a bench as shown in Fig. 4.17. The height of the bench should not be less than 3 – 4 m, and further benches should be introduced where the bench height exceeds about 10 m. A typical operating cycle is shown in Table 4.1.

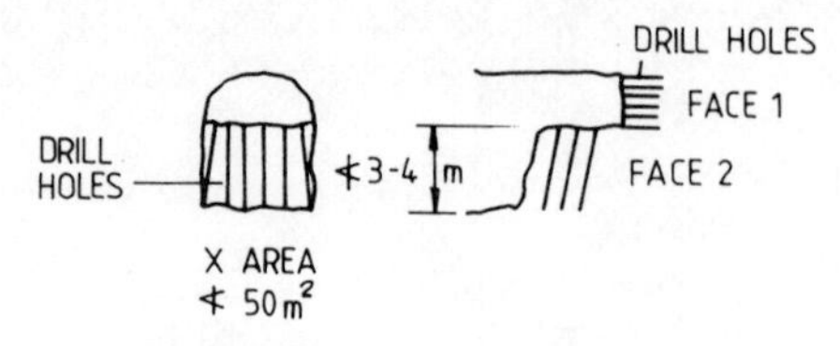

Fig. 4.17 Use of benching in tunnelling

Table 4.1 Time cycle in hours per blasting sequence

	Hours
Boring, charging, blasting	3.0
Ventilation pause	0.25
Loading-transport	3.25
	6.50
Rock bolting	1–4

Blasting data (Leins (1972))

Benching

Operation	Drilling required (m) per m^3 of rock blasted		
	Soft	Medium	Hard
Quarrying	0.3	0.4	0.5
Large excavations	0.7	1.1	1.5
Narrow excavations	1.0	1.5	2.0

Tunnelling

Figures 4.18 and 4.19 illustrate comparable output data for tunnelling in hard rock, where it can be seen that for small diameter tunnels considerably more drilling is required than for large diameter sections or in normal benching. This is mostly due to the degree of fixity of the burden, and the need to open up from a single free face cut section for each and every advance.

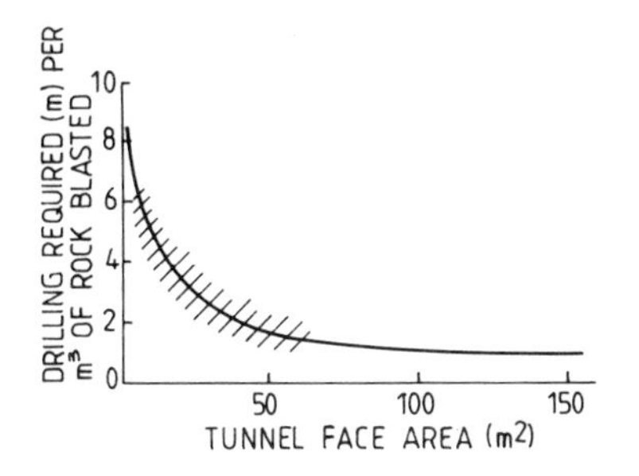

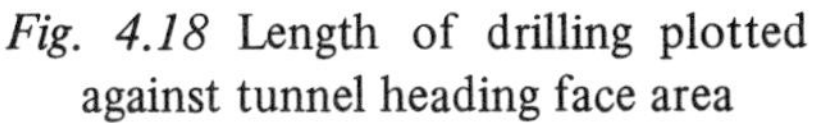

Fig. 4.18 Length of drilling plotted against tunnel heading face area

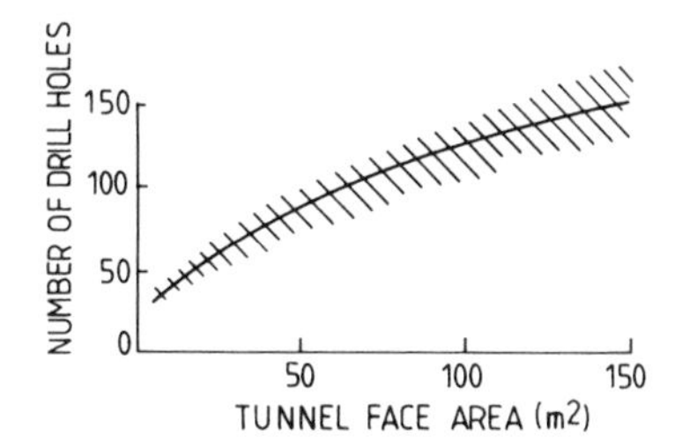

Fig. 4.19 Number of drill holes plotted against tunnel heading face area

Consumption of explosive (approximate)

Benching

Operation	High explosive (kg) per m³ of rock blasted		
	Soft	Medium	Hard
Quarrying	0.1	0.2	0.3
Large excavations	0.3	0.45	0.6
Narrow excavations	1.0	1.5	2.5

Tunnelling

Figure 4.20 illustrates comparable explosives data for tunnelling in hard rock.

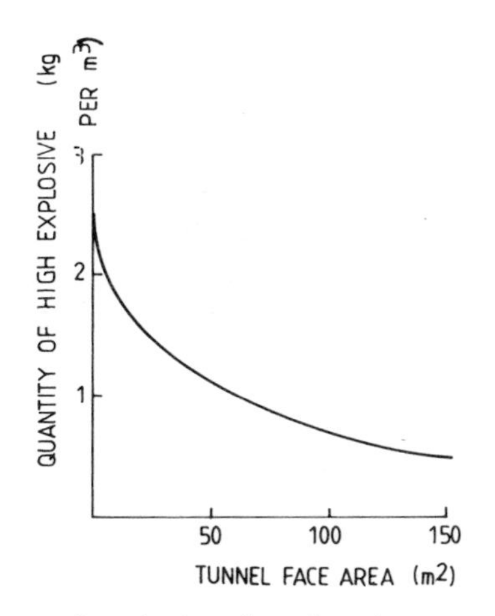

Fig. 4.20 Quantity of explosive plotted against tunnel heading face area

Underwater explosives

Water-resistant explosives are necessary and because of the difficult working conditions, e.g. oversize boring, poor packing of explosive, etc., energy loss occurs, resulting in a requirement of 150 to 400% more explosive than for blasting on land.

Types of explosive

An explosive charge on detonation reacts to form a large volume of gas at high temperature. The release of the gas is almost instantaneous and produces high pressure and shock waves (see earlier description on page 51) which continues until the reaction is complete.

Many types of explosive of varying strength, detonation times, etc., have been developed and they may be classified into:

(i) slow explosives
(ii) high or quick explosives
(iii) initiating explosives.

(i) Slow explosives

The most common type is known as black powder, which consists of a mixture of saltpetre (potassium nitrate), charcoal and sulphur.

Black powder can be ignited by explosure to an open flame and must be closely confined to maximise the effect of the relatively low gas pressure produced. The velocity of detonation is slow, about 600 m/s and thus there is little crushing of the rock near the centre of the explosion. The main effect is a bursting action causing displacement and fragmentation as the gas is slowly released from the reaction. Its uses are mainly limited to quarrying and in safety fuse cord.

(ii) High explosives

These are not very sensitive to detonation by spark or flame and require initiating explosives. When detonated in this way, the reaction takes place at up to 9000 m/s with accompanying high gas pressure. The surrounding rock is thus crushed, cracked and shattered. This type of explosive is therefore ideally suited to quarrying and construction applications. The most common types of high explosive are:

(a) nitroglycerine
(b) ammonium nitrate plus fuel oil (AN/FO)
(c) slurries

They have the following properties:

(i) density – approx. 1.5 g/cm^3 for gelatins, 1.1 for powders.
(ii) specific energy 1000 –2000 kcal/kg.

(a) *Nitroglycerine*
Is produced by the action of nitric and sulphuric acids on glycerine. The resulting mixture is highly sensitive to impact and temperature and can be detonated by the action of a gentle blow. Inert substances such as charcoal, sodium nitrate or nitrocellulose may be added to improve stability but these generally reduce the weight strength of the explosive. Nitroglycerine explosive compounds are often referred to as Dynamite when composed of glycerine-ethylene, glycol nitrates, ammonium nitrate, wood chippings and inert substances. The compound has a dry, granular consistency. Dynamites for use in water are best gelatinised; blasting gelatin containing a mixture of nitroglycerine and nitrocellulose was formerly popular although more powerful gelatins are now available.

(b) *Ammonium nitrate*
Nitroglycerine-based explosives are fairly expensive and a mixture of ammonium nitrate and oil (AN/FO) provides an alternative substance. The compound is mostly used in powder form and can be poured directly into the drill hole. However a powder loader using compressed-air greatly speeds up the operation. AN/FO is not water-resistant, and therefore is not suitable for use in wet drill holes, but this may be overcome by using a gelatinised form, possibly containing a nitroglycerine additive.

The gelatins are usually manufactured in paper or cardboard cartridges for ease of handling and loading.

(c) *Slurries*
These consist of aluminium particles, or TNT (trinitrotoluene), plus an inert stabilising compound, all suspended in water. This type of explosive has a slightly greater weight strength than AN/FO and can be used in wet conditions, but it is slightly more expensive.

(iii) Initiating explosives

Because high explosives are not very sensitive to detonation by spark or flame, an initiating explosive is required to set off the charge. These are extremely sensitive and easily exploded, producing sufficient shock and temperature rise to induce a reaction in the high explosive. Such explosives are used in small quantities to form detonators.

Detonation of explosives

The initiating explosive in a detonator may be ignited either with a safety fuse cord (Fig. 4.21) or electrically (Fig. 4.22).

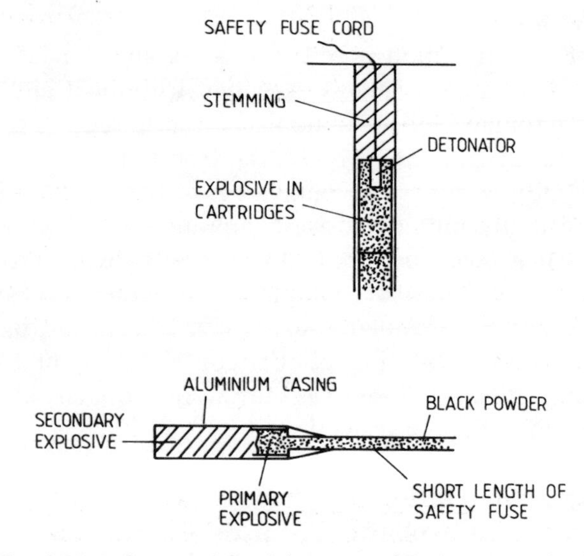

Fig. 4.21 Safety-cord fired detonator (black powder fuse)

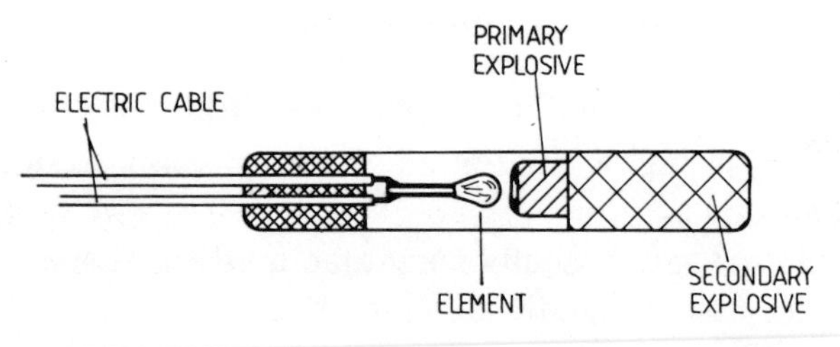

Fig. 4.22 Electrically fired detonator

Safety Fuse cord

This has a black powder core sheathed in plastic. The cord is designed to burn at a rate of approximately 100 m/s and a short length is led to the detonating cap which is embedded in the high explosive. When black powder is itself the explosive compound, then the fuse cord is sufficient to initiate the charge and a detonator is not required. Although safety fuse cord can be ignited by flame, it is more usual when several shots are to be fired to link the free end of each safety fuse cord to igniter cord, which is then led back to a safe location. Igniter cord can be obtained with slow (3.5 m/s), moderate (33 m/s) or fast (50 m/s) burning rates.

Detonating caps (blasting caps)

(i) *Non-electric* (*Fig. 4.21*) - comprises an aluminium tube containing a powerful initiating explosive, such as pentaerythritol, and a priming charge of lead oxide. Caps are manufactured in different strengths, the common sizes being No. 6 and No. 8, which give efficient detonation with most types of blasting explosives.

With this type of detonator, time delays are introduced by varying the length of the safety fuse cord.

(ii) *Electric detonators* (*Figs. 4.22 and 4.23*) - are also made in varying strengths, with the initiating explosives contained in an aluminium or copper casing (Fig. 4.22). The element in the cap is simply connected by wires to an electric source, which subsequently sets off the primary explosive. Delays ranging from about 8 to 8000 milliseconds can be incorporated (Fig. 4.23).

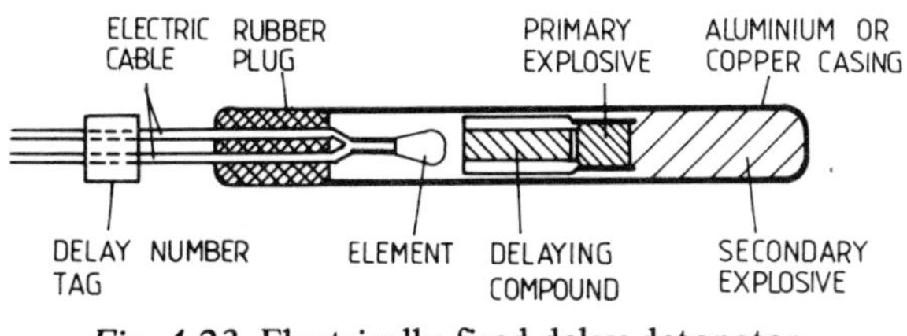

Fig. 4.23 Electrically fired delay detonator

Circuit design

In simultaneous blasting the caps can be arranged in series as shown in Fig. 4.24. But for delayed charges, obviously the circuit must be

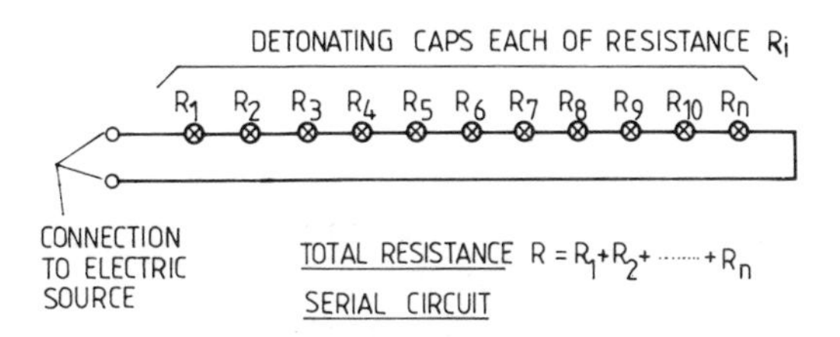

Fig. 4.24 Serial row detonator arrangement

maintained after the first shot, and therefore a parallel circuit is required. Figure 4.25 illustrates an arrangement for single-row delays and Fig. 4.26 for multiple-row delays. The R_i values indicate the resistance in ohms of each cap and with a knowledge of the current (I)

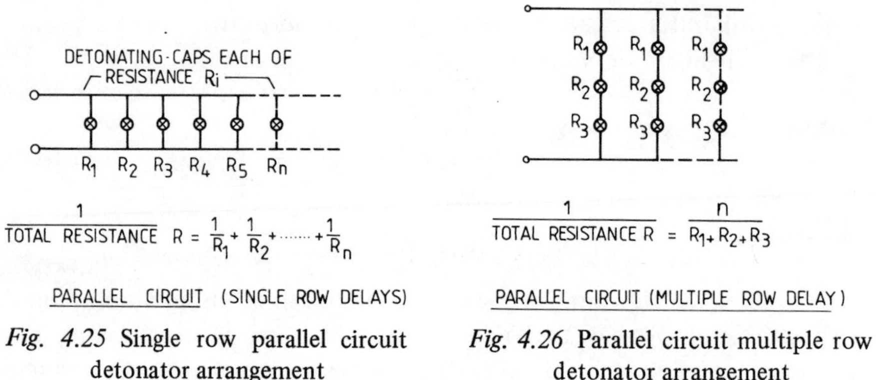

Fig. 4.25 Single row parallel circuit detonator arrangement

Fig. 4.26 Parallel circuit multiple row detonator arrangement

needed to fire each cap (about 1.5 amps) the total volts can be calculated from the formula $V = IR_{\text{total}}$. An exploder (source power) of the required size is connected to the other ends of the electric wires.

Detonating fuse

As an alternative to the use of safety fuse cord and its associated detonating cap, detonating fuse can be used to fire the main explosive. It consists of a central core of pentaerythritol-tetranitrate explosive wrapped in tape. The outer casing is made up from strands of yarn and sheathed in plastic (Fig. 4.27). This type of fuse cord produces a high

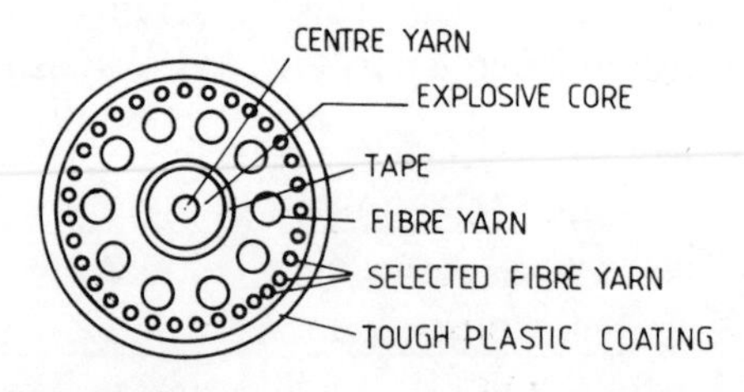

Fig. 4.27 Detonating Fuse (high explosive fuse)

velocity of detonation (6000 m/s) and must itself be initiated by a detonating cap. It can be used for firing simultaneous charges (Fig. 4.28) or with delays by the insertion of detonating relays (Fig. 4.29). Because the detonating cord can be run alongside or through the high explosive cartridges or powder in the drill hole (Fig. 4.30) there is less risk of unscheduled ignition than with a single blasting cap.

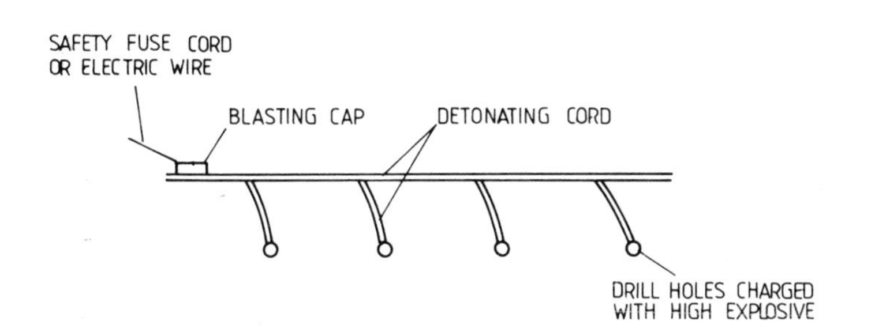

Fig. 4.28 Simultaneous blasting arrangement with detonating fuse

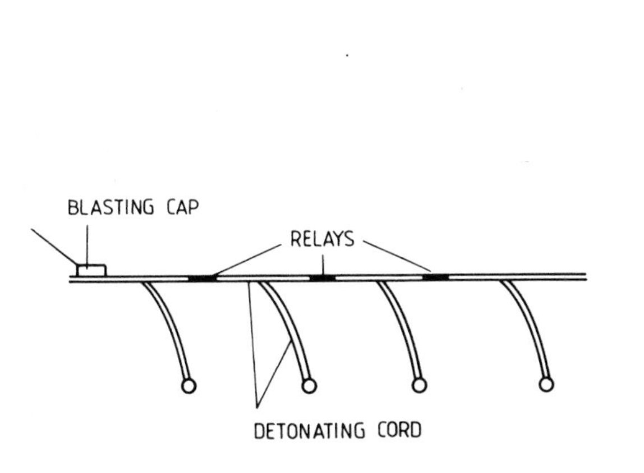

Fig. 4.29 Delay blasting arrangement with detonating fuse

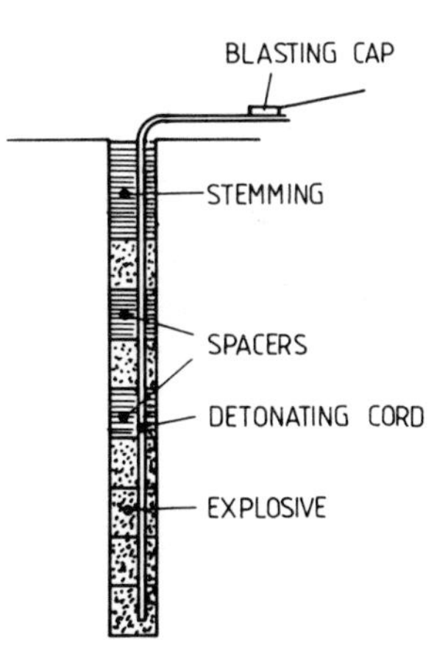

Fig. 4.30 Distribution of explosive and detonating fuse

Loading the drill hole

The hole should first be cleaned out with compressed-air. When using powder explosive, pneumatic equipment can speed up the loading operation, but unless special equipment is available, explosive supplied in cartridges must be laboriously loaded into the drill hole and pressed firmly down with a long pole, the objective being to produce as high a density of packing as possible (P).

The top part of the drill hole should be plugged with damp sand or clay, which acts to confine the energy of the explosion. Ideally the length of stemming should not be less than the width of the burden.

Vibrations from blasting

The U.S. Bureau of Mines (USBM) recommends that vibration levels resulting from blasting operations be limited below a peak particle velocity of 51 mm/s if structural damage is to be avoided. Several researchers have published results of the effects of vibrations and these are summarised in Table 4.2.

Table 4.2 Structural damage as related to peak particle velocity

Peak particle velocity mm/s	Langefors (Sweden)	Edwards (Canada)	Bureau of Mines (U.S.A.)
250	Serious cracks	Damage	Major damage (Fall of plaster, serious cracking)
200	Cracks	Damage	Minor damage (Fine plaster cracks, opening of old cracks)
150	Small cracks and fall of loose plaster	Damage	Minor damage (Fine plaster cracks, opening of old cracks)
100	Caution	Caution	Caution
75	Caution	Caution	Caution
50	No noticeable cracks	Caution	Caution
25	No noticeable cracks	Caution	Caution
00		Safe limit	Safe limit

A useful formula for estimating the likely particle velocity resulting from an explosive charge is given by USBM as

$$V = k \frac{D}{\sqrt{E}} - B$$

where E = explosive charge (kg)
V = particle velocity (mm/s)
D = distance from shot to recording stations (m)
k and B are empirical co-efficients which can be determined from field trials with instrumentation investigations

Air blast

If part of the explosion escapes to the air, the expanding gases push back the air causing a local increase in air pressure.

Windows can be broken when this over-pressure lies in the range of 0.05 to 0.14 kg/cm^2.

References

Atlas Copco AB (1961) Manual on Rock Blasting, 8 (20). Stockholm.

Bauer, A. and Calder, P. N. (1974) 'Trends in explosives, drilling, and blasting'. *Canadian Min. and Met. Bull.* **67**, Feb.

BS 6187 (1982) *Demolition.* British Standards Institution (formerly CP94).

BS 5607 (1978) *Safe use of Explosives in the Construction Industry*, British Standards Institution.

Grimshaw, G.B. (1978) 'Pre-splitting and smooth blasting are improved methods'. *Mining Equipment International.* Jan./Feb.

Harris, C.M. and Crede, C.E. (1961) *Shock and Vibration Handbook.* Vol. **I–III**, McGraw-Hill, N.Y.

Henrych, J. (1979) *The Dynamics of Explosion and its Use.* Elsevier, Holland. (Translated from the Czech by Major, R.)

Johansson, C.H. and Lundborg, N. (1958) 'Firing electrically'. *Manual on Rock Blasting.* **16** (10), Atlas Copco, Stockholm.

Langefors, U. and Kihlstroem, B. (1963) *The Modern Technique of Rock Blasting.* Wiley & Son Inc., New York and Almquist and Wiksell, Stockholm.

Leins, W. (1972) *Tunnelbau.* Aachen Technische Hochschule, Germany.

Livingston, C.W. (1960) *Explosive in Ice.* U.S. Army, Snow, Ice and Permafrost Research Establishment, Corps of Engineers, Wilmette, Illinois.

McFarland, D.M. and Rolland, G.F. (1965) *Handbook of Electric Blasting.* Wilmington, Del. Atlas Chemical Industries Inc.

Nobel's Explosives Co., (1972) ICI Blasting Practice. Scotland, 4th Edn.

Urbanski, T. (1981) *Chemistry and Technology of Explosives.* Pergamon.

Wiss, J.F. (1974) 'Vibrations during construction operations'. *Journal of the Construction Division*, ASCE, Sept.

CHAPTER FIVE

Soft ground drilling

Boreholes are necessary in soft ground for:

(i) soils investigation
(ii) ground anchorages
(iii) foundations, piles, both in situ and precast
(iv) water wells
(v) horizontal conduits
(vi) piled walls

The two principal methods for drilling holes are rotary, or auger, boring and conventional grabbing with a bucket. The choice of type depends upon the ground conditions, the diameter and depth of borehole required and the cost and availability of the equipment. A summary of the capabilities of the various methods is shown in Table 5.1.

Table 5.1 Methods of producing boreholes in soft ground

Method	Max. depth (m)	Max. dia. (mm)	Soil type
Soil investigation boring	When bedrock is reached	150-200	All
Continuous flight augering	30	Up to 600	Firm uniform soils
Intermittent flight augering	50	Up to 2500	Firm uniform soils
Rotary boring with buckets	50	Up to 1500	Free-flowing soils
Rotary boring with belling buckets	30-40	Up to 6000	Cohesive soils
Grabbing	100	500-2000	Difficult soils and those containing small boulders
Circulation drilling	100	300-2000	Soft rock

Soils investigation techniques

In civil engineering and building, soils investigation is often undertaken to determine the suitability of the soil to support foundations; the size of, and depth to which to sink or drive piles; groundwater conditions, etc.

Boreholes for sampling and soil testing must be large enough to accommodate the sampling equipment, and 150 –200 mm diameter is usually sufficient to do this. The borehole is generally terminated on reaching bedrock. Up to this stage the methods used include:

(i) shell and auger boring
(ii) percussive drilling
(iii) wash boring

Simple equipment is necessary, which can be quickly set up, transported, and operated by at most 2 men.

(i) Shell and auger boring

The machine comprises a three- or four-legged derrick, sometimes fitted with detachable wheels to facilitate transportation, and a winch and clutch capable of lifting about 4000 kg. The height from ground to centre line of sheave is approximately 5 m with an overall weight of 1000 kg (Fig. 5.1).

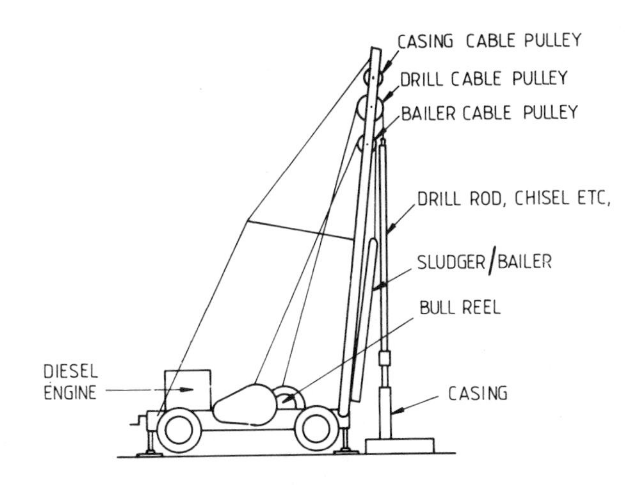

Fig. 5.1 Soils investigation rig

(a) *The shell cutter*

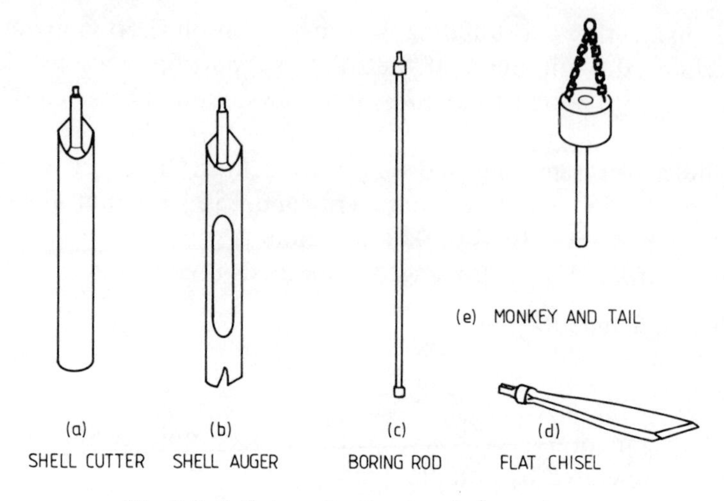

Fig. 5.2 Soils investigation borehole equipment

The shell cutter shown in Fig. 5.2(a) consists of a steel tube fitted with a cutting edge, which is hung from the rig and allowed to fall. A core is cut which is retained in the cylinder. The cutter is then raised and the material removed. As the process is repeated, the borehole is gradually formed. In sandy soils a small clack valve is incorporated into the cylinder to prevent the material flowing out.

(b) *The auger*

For shallow work (4 - 5 m max. depth) an auger is often preferred (Fig. 5.2(b)). This consists of a hollow tube, connected to boring rods, which is rotated manually at ground level. The soil is removed from openings in the side of the cylinder wall.

The auger is used for coring out clay and other fairly soft strata.

The advantages of these systems are that the equipment is simple to erect and use and that undisturbed samples can be obtained, thus relieving the operator of having to identify the strata in the field.

(ii) Percussive drilling

During the investigation a thin rock layer or small boulder may be encountered. The shell or auger is then replaced by a chisel of approximately 680 –700 kg weight (Fig. 5.2(d)). The continuous pounding by this chisel breaks up the rock. The cuttings and slurry are removed by baler or shell cutter. The process is applied until the operator is satisfied that the particular stratum is in fact bedrock or otherwise.

(iii) Wash boring (*Fig. 5.3*)

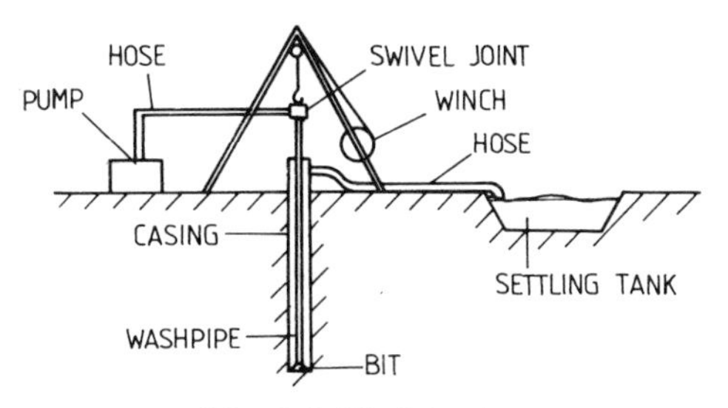

Fig. 5.3 Wash boring

The wash boring method uses a light tripod, a small winch and a small pump, and sometimes a small power-driven capstan is provided.

Casing for the borehole is first driven into the ground. The soil is then removed from inside the casing by chiselling with the hollow drill pipe supported by the winch. The soil is loosened by a combination of the chiselling action and the stream of water or drilling mud issuing from the lower end of the wash pipe.

The loosened debris is carried up the annular space between the casing and wash pipe into a settling tank.

Unfortunately the samples are disturbed samples and are therefore suitable only for identification of the various soil strata encountered i.e. unsuitable for laboratory testing.

The main advantage of this method is that the equipment is easy to assemble and operate. However, it can only be used to good effect in softer type soils, such as sands and clays.

Although the equipment is simple, it requires an experienced operator to interpret the debris from the borings. The method is also very slow.

An even cruder alternative is Wash probing, (Fig. 5.4), which gives a

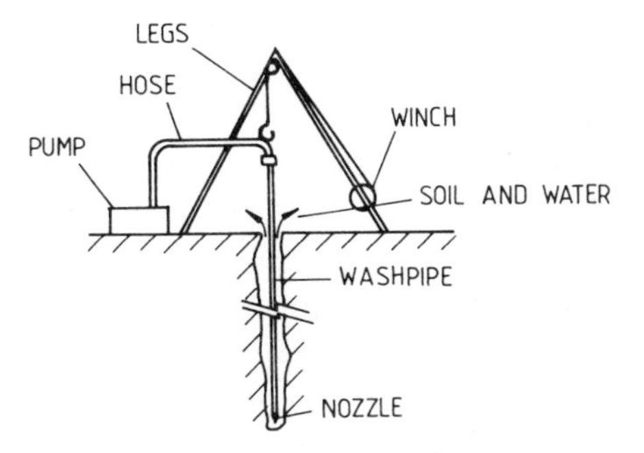

Fig. 5.4 Wash probing

rough indication of the position of the change between a soft or loose soil and that of a more compact soil. The casing is not necessary, the change being determined by the operator from the feel of the wash pipe.

Lining tubes

In most non-cohesive soils the borehole will probably need a lining. This consists of sections of tubular casings driven from the surface by means of a 'monkey and tail' operated from a winch (Fig. 5.2(e)).

Continuous flight augering

The rotary boring machine designed for drilling rock, described in Chapter 3, can also be fitted with a continuous flight auger (Fig. 5.5) and bits suitable for boring in soft ground. Flushing is not required.

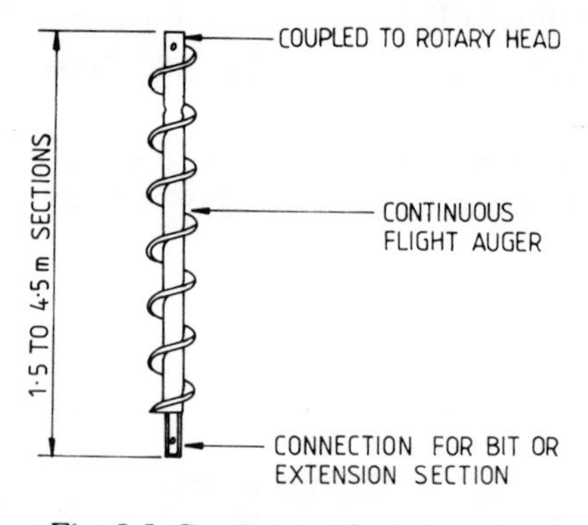

Fig. 5.5 Continuous flight augering

The spiral flight on the auger has a pitch designed to bring the spoil to the surface. By inserting spiral sections, boreholes of up to 600 mm diameter can generally be sunk to depths of 30 m. However, depending upon the hoist capacity, depths of up to 100 m with smaller diameters have been obtained.

This method is best suited to boulder-free foundations, such as sand, gravel, clay, slate, chalk, limestone and coal. For harder foundations, the spiral auger is removed and the rig is converted to operate with either roller or drag bits (see Chapter 3) as a conventional rotary drill.

The method is unsuitable where free-flowing materials are encountered which cannot be retained on the flighting, such as water-bearing gravels, sands and silts. For these formations a grabbing or rotary bucket technique is usually more appropriate. A lining tube is used where the sides of the borehole are likely to collapse.

Typical uses of this method are for sub-surface exploration, well drilling and boreholes for ground anchors.

It should be noted that the diamond drilling rig is unsuitable, because of its high rotation speeds and relatively low torque availability.

Augers (*Fig. 5.5*)

The augers for continuous augering are generally made in 1.5, 3, 4.5 and 6 m sections and are added as the drilling progresses.

Drilling heads

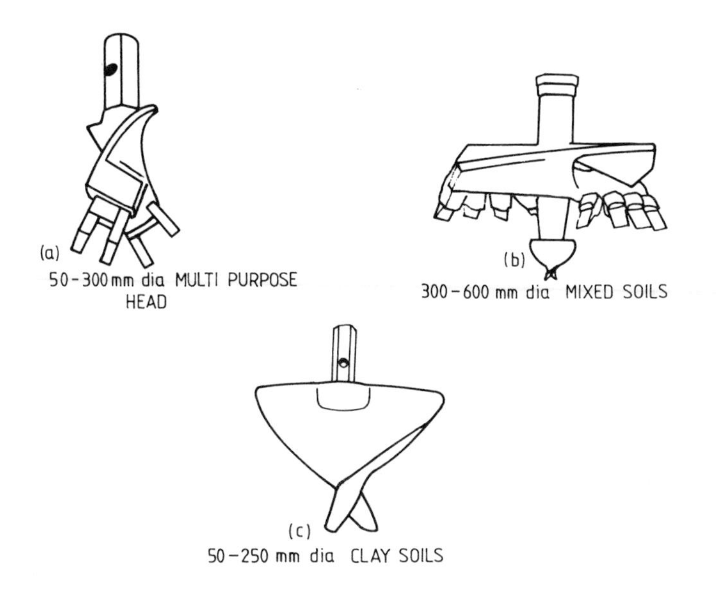

Fig. 5.6 Auger bits for soft to medium soils

For common earth, sand and some gravels, a finger bit (Fig. 5.6(a)) of up to 300 mm diameter is suitable, but in clay soils the specially-designed clay bit is preferred (Fig. 5.6(c)). For larger diameters the earth bit is more appropriate (Fig. 5.6(b)). The method can also be used in frozen soils and soft to medium hard rocks, with rock augers (Fig. 5.7) but an increased feed force may be necessary to produce adequate penetration rates.

Intermittent augering

While the continuous flight auger has adequately serviced the need for relatively shallow and small diameter boreholes, the demand for large

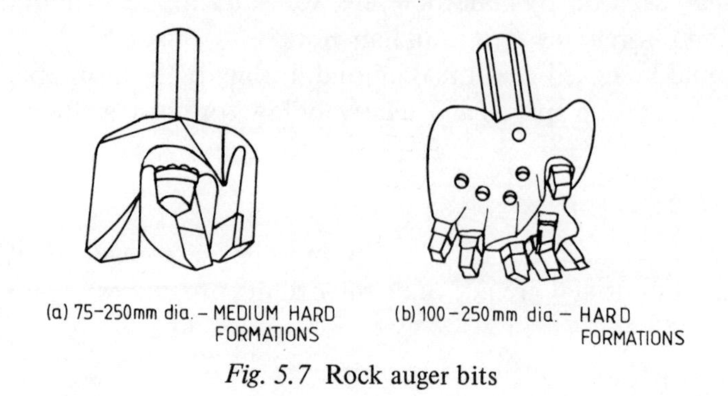

Fig. 5.7 Rock auger bits

diameter borings, especially for foundation piles, has required a modified augering procedure. The intermittent method uses a short flight auger or drilling bucket and holes of up to approximately 2.5 m diameter can be drilled, to depths of 50 m.

Equipment

Drilling rig

The design and construction of the rig is similar to the rotary drilling machine used for rock boring, but no flushing medium is needed. The basic machine consists of either a truck (Fig. 5.8) or crawler crane (Fig. 5.9) supporting a mast, telescoping kelly bar and drill head;

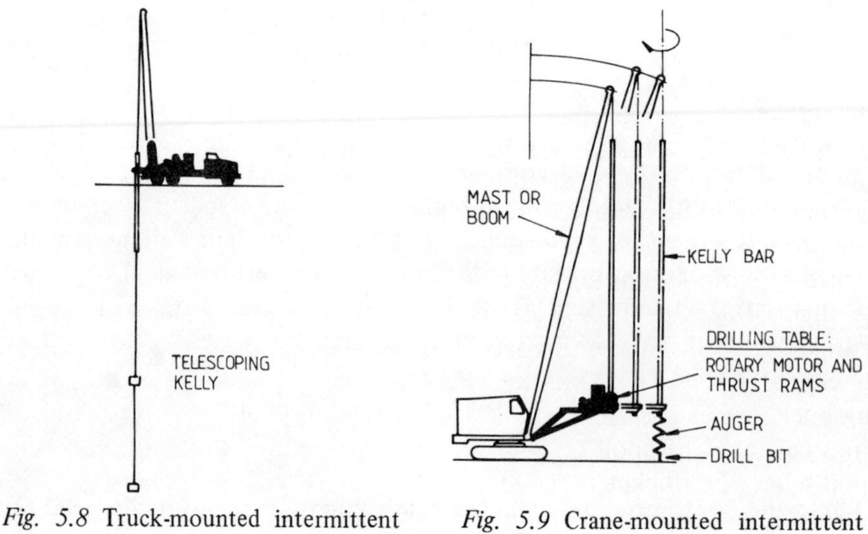

Fig. 5.8 Truck-mounted intermittent auger rig

Fig. 5.9 Crane-mounted intermittent auger rig

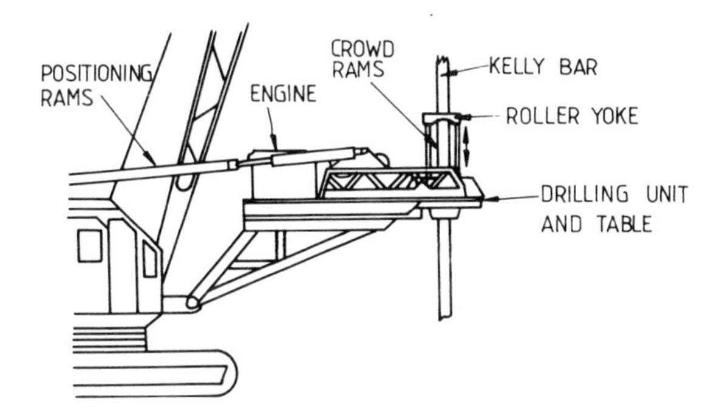

Fig. 5.10 Drilling unit for intermittent augering

a drilling table and a separate diesel engine to power the drilling units.

The details of the arrangement are shown in Fig. 5.10. The mast guides the telescoping kelly bar, the latter being attached to the crane or truck base through two sets of hydraulic rams, and a base unit connected to the crane slewing ring. Two ropes are used, one to operate the kelly bar, the other to handle any casing tubes. Power is supplied from a separate diesel engine to drive a hydraulic system which provides the turning movement and crowding force at the turntable. High rotation speeds of up to 200 r.p.m. are possible, depending upon the torque and gear selected, and depths down to 100 m have been achieved, but more usually the telescoping action of the kelly is available only up to 50 m depth, with additional kellies having to be added manually thereafter.

Auger and drill bit

This method is suitable for non conglomerate soils, and drill bits similar to the continuous flight augering method are used. However, unlike the continuous auger, the intermittent method requires only a short section of flighting above the drill bit (Fig. 5.11). The kelly bar must therefore be raised to ground level when the auger is fully laden. The material is discharged by simply spinning the soil off the flighting.

The flight auger can bore holes of up to about 2.5 m diameter but unfortunately is unsuitable for free-flowing materials and mud, etc., which cannot be retained on the flighting. In such circumstances a drilling bucket is necessary (Fig. 5.12). This has a conical base with two openings to allow the spoil to enter. Flaps cover the openings to retain the spoil when the bucket is raised from the hole. The bucket is fitted with two sets of digging teeth mounted across the base, and holes of up to approximately 2 m diameter can be bored.

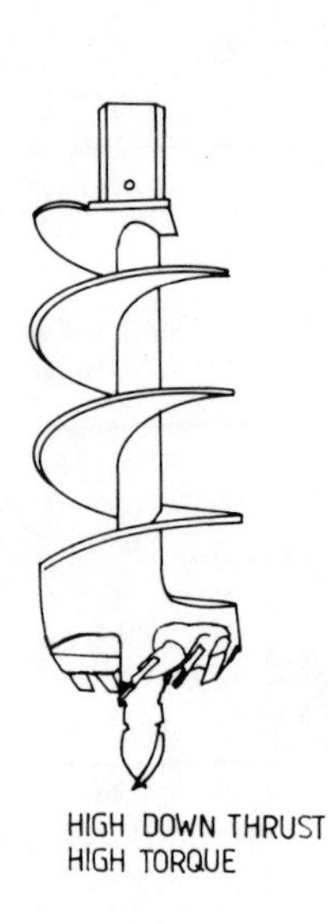

Fig. 5.11 Intermittent flight auger and bit

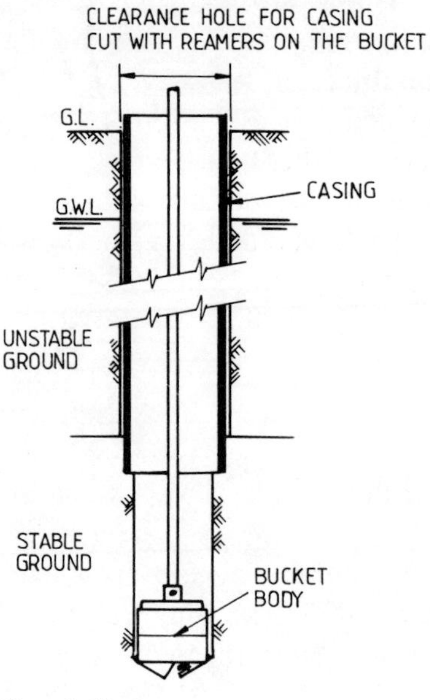

Fig. 5.12 Intermittent bucket auger

Belling buckets

In stable soils, the diameter of a borehole produced by a rotary method can be extended to about three times that of the hole, with special purpose belling tools.

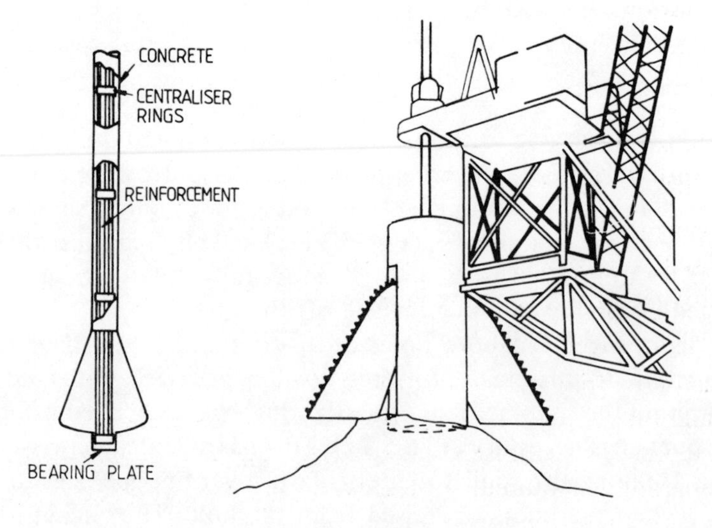

Fig. 5.13 Under-reamed borehole and belling equipment

A hole is drilled in the normal way with an auger or bucket, which is then replaced by the belling bucket (Fig. 5.13). With this bucket standing on the bottom of the hole, the belling arms are gradually opened out as the bucket is rotated. However, there is a tendency for part of the belled roof to fall away before concrete can be placed (e.g. in situ piling), and therefore good practice requires an inspector to be lowered down the borehole in a safety skip (Fig. 5.14) and the work examined, cleaned up, etc. before concrete placing begins.

Fig. 5.14 Inspecting an under-reamed borehole

Borehole lining

In many situations the ground will be stable and self-supporting while drilling takes place. However, where the ground would collapse or where inflowing water would cause problems of stability, the sides of the borehole require supporting. This may be achieved by several techniques, such as casing or fluid pressure methods. In general, casing is preferred, as shown in Fig. 5.12. In non cohesive soil the casing may be installed and extracted with a vibrating method described for sheet piling (page 160). Alternatively the casing may be screwed in by:

(i) using the kelly
(ii) oscillating with a special hydraulic oscillator, or
(iii) driving with a drop hammer suspended from the mast of the crane or rig

Because of the limited headroom under the drilling table, it is usual to install the casing in sections which are progressively screwed together as the borehole depth is extended. Often a clearance hole can be drilled before casing is required, thus allowing the lead casing to be of greater length than the extension sections.

Table 5.2 Rotary boring rig data (telescoping type)

Auger or bucket diameter (m)	Max.* drilling depth (m)	Power Unit (kW)	Max. r.p.m.	Crowd or feed force (kN)	Torque (kN.m)	Equivalent crane size (tonnes)
0.7	30	70	200	120	17	15
0.9	30	80	160	120	25	20
1.5	35	90	130	130	35	30
2.0	40	110	105	150	55	40
2.5	45	160	70	150	120	50

Output 10–20 m per hour. Accuracy 1 in 75 vertically.
*Can be extended to approximately 100 m with additional drill stems.

Characteristics of the auger boring method

Torque

The main characteristic of this type of rotary boring method compared to small diameter rotary rock boring is the need to generate very high torque because of the large diameter boreholes involved. Indeed, as an auger cuts through the earth, the torque required to keep it rotating varies continuously, as illustrated by the example shown in Fig. 5.15. Clearly it is therefore highly desirable to have sufficient power in reserve to respond to a rapidly changing torque need.

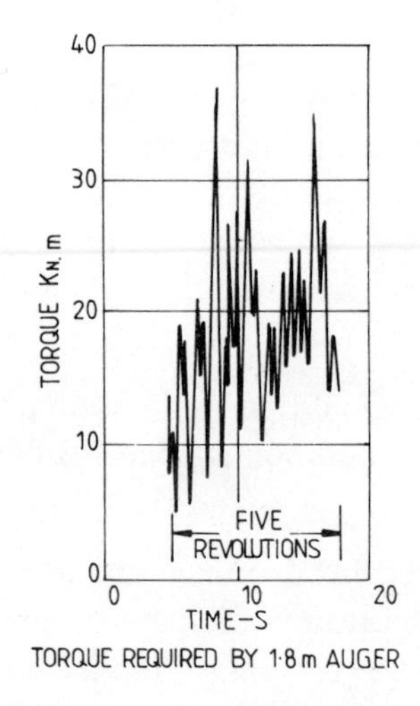

Fig. 5.15 Torque required for auger boring

Crowd and hoist

The weight of the kelly bar provides some downward thrust but for work in medium to hard soils, additional down-force must be provided by the hydraulic feed.

Grabbing methods

When poor ground, or obstructions are frequently encountered in a soil such as a conglomerate, then the most economical solution may be to dispense with the rotary system and concentrate on a grabbing or chiselling technique, such as the Benoto method. The Benoto machine consists of a basic frame supporting a mast, which houses a grabbing bucket, engine and winches (Fig. 5.16). The borehole casing is both oscillated and forced downwards, using hydraulic rams (Fig. 5.17).

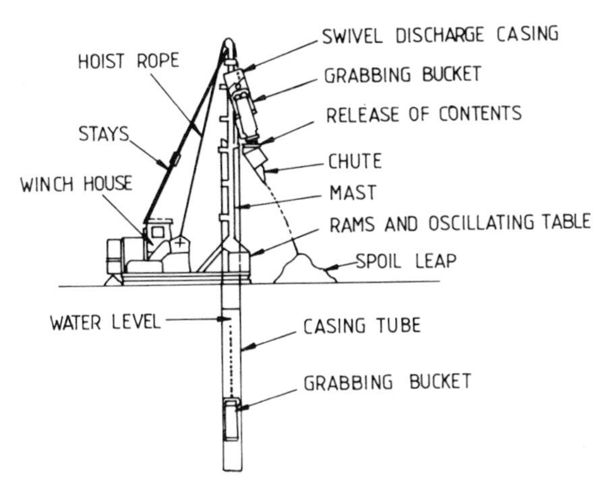

Fig. 5.16 Borehole drilling by grabbing action

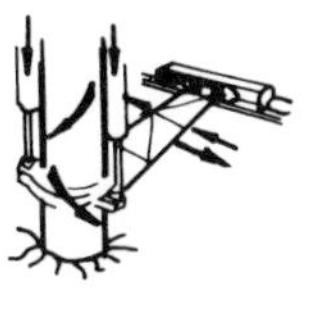

Fig. 5.17 Installing borehole lining with an oscillating-type rig

In soft soils the lining is pressed into the soil ahead of the grab, while in harder materials the casing must first be under-reamed. Like the casing in the augering method, sections of pipe, 2, 4 or 6 m long, are screwed together. Boreholes from 0.5 to 2 m diameter can be obtained and rakes of up to 12° from the vertical are possible. Depths of 100 m are achievable with an accuracy of up to 1 in 200 vertically.

The grabbing bucket (*Fig. 5.18*)

The grab is controlled by a single line from the winch (or alternatively

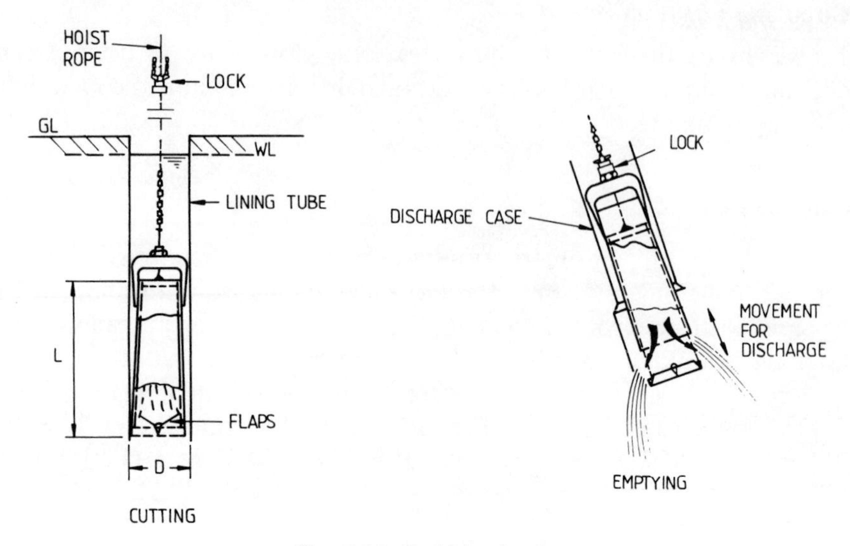

Fig. 5.18 Grabbing bucket

by a hydraulic kelly bar). A simple device on the line opens the bucket for discharge of the contents. Typical bucket sizes are shown in Table 5.3. The rate of excavation is slow and depends upon the material and size of grab. The method is suitable for most types of material and can even be used as a rock chisel. In soft to medium soil, a 1 m^3 capacity grabbing bucket could excavate 2 – 5 m^3 per hour.

Table 5.3 Rope grab data (*see Fig. 5.18*)

D (mm)	L (mm)	Weight (kg)
350	1500	450
450	1500	450
550	1750	450
650	1750	450
800	2500	1000
950	3200	1700
1200	3800	2500
1500	5300	7250

The Benoto rig

Total weight	30 tonnes
Engine	100 kW
Angle boring	Up to 12°

Rigs of a similar type include the Bade Hochstrasser-Weise and ACE machines.

Circulation drilling (*Fig. 5.19*)

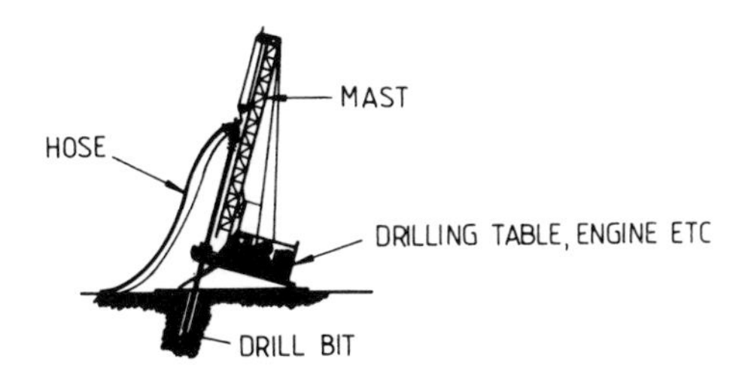

Fig. 5.19 Circulation drilling

The installation of lining tubes in very deep borings may be technically difficult and very expensive. This problem may be overcome by introducing fluid pressure into the borehole to stem the flow of incoming water and to support the soil.

There are three basic circulation drilling methods:

(i) direct circulation
(ii) reverse circulation
(iii) reverse circulation with air lift

Reverse circulation is best suited to large diameter boreholes.

(i) Direct circulation (*Fig. 5.20*)

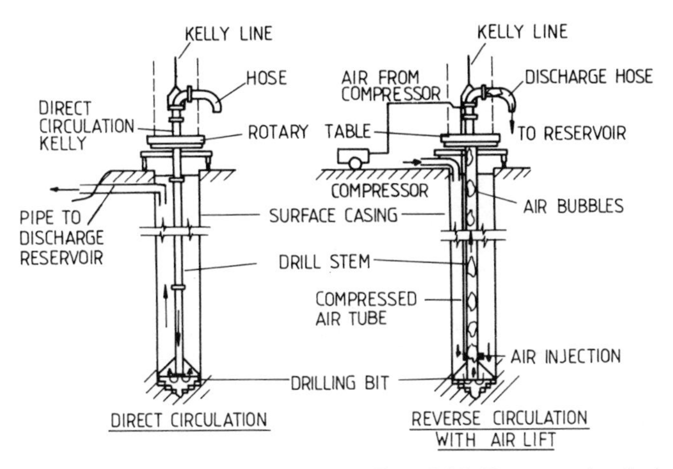

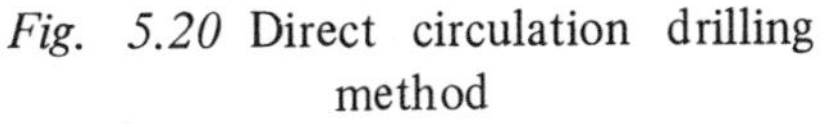

Fig. 5.20 Direct circulation drilling method

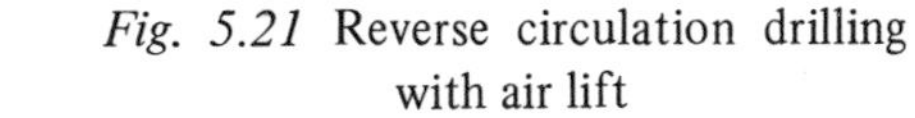

Fig. 5.21 Reverse circulation drilling with air lift

Drilling fluid is pumped down a hollow drill pipe, around the drill bit and back to the surface in the annular space around the drill pipe. Cuttings are carried to the surface by the flow.

The volume of drilling fluid necessary to give a satisfactory raising velocity within the annular space depends on the volume of that space. This, in turn, determines the capacity of the pump needed (see Flushing, page 41).

(ii) Reverse circulation (*Fig. 5.22*)

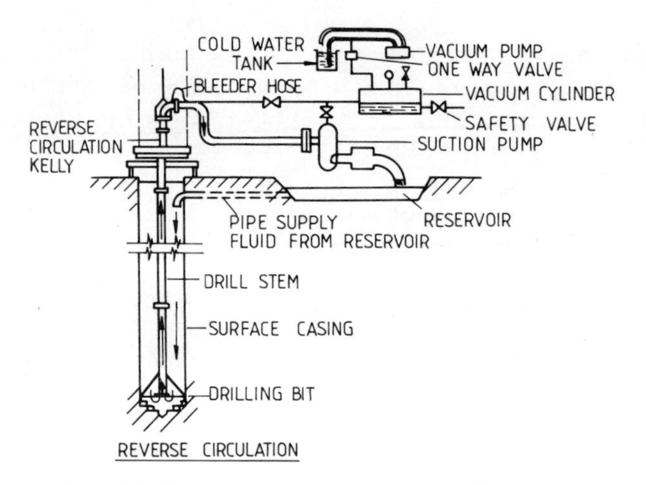

Fig. 5.22 Reverse circulation with vacuum pump

In this counter-flow system, drilling fluid is introduced into the annular space around the drill pipe where it moves down and around or through the drill bit. The fluid – along with excavated material – is then drawn back up to the surface through the inside of the drill pipe by a high capacity, low head, centrifugal pump. This system must be primed with a vacuum pump and the hole kept full of fluid at all times.

(iii) Reverse circulation with air lift (*Fig. 5.21*)

Drilling fluid is introduced into the bore hole as with the pumped reverse circulation method. The driving force to return the flow, and cuttings, to the surface is created by injecting air into the drill string. The composite mixture of fluid and air bubbles, which is lighter in weight than the fluid outside the drill pipe, forms a pressure differential which creates the flow.

An adequate fluid level must be maintained in the hole to provide the necessary hydrostatic head.

This method is simple and robust and requires no pumps. Much deeper boreholes can be obtained by this method than by direct or reverse circulation. Approximate consumption of compressed-air is

Pipe diameter (mm)	*Free air* (m^3/min)
150	4.5 to 6
200	6 to 10
300	15 to 20

Circulation fluid

In sandy soils the interstices may be partially sealed and the grains stabilised by adding bentonite to the pump water, to produce a thixotropic solution weighing approximately 1200 – 1400 kg/m^3. Usually a 4% solution by volume is sufficient but up to 10% may be required in very difficult conditions.

Concrete can be deposited in the bentonite through a tremie tube and the displaced fluid drawn off. Sometimes it may be necessary to install a short section of lining casing, especially when drilling over water, to give a firm support, near the top of the borehole.

Rig data

Engine	Up to 100 kW
Rotation speeds	Up to 60 r.p.m.
Circulation rates	Up to 8000 l/m
Torque	Up to 200 kN.m.

This method is principally designed for deep boreholes of 300 – 2000 mm diameter, in uniform soils and consolidated materials. The bit used is similar to that required for a drilling bucket, with two or three rows of teeth mounted across a circular ring (Fig. 5.23).

Outputs (i.e. rate of penetration)

Common earth	3 – 4 m/h
Sands and clays	2 – 3 m/h
Gravel	1.5 m/h

2 man operation.

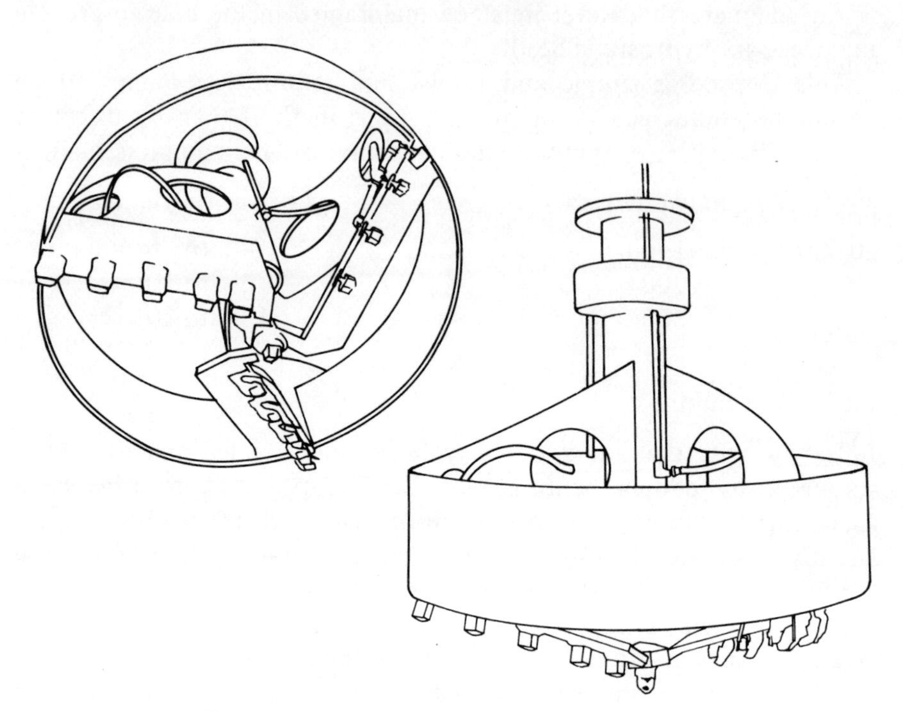

Fig. 5.23 Cutting head for soft to medium soils

Drilling rock

Although not a common feature in civil engineering work, roller bits (Fig. 5.24) have been developed to sink mine shafts through rock strata. The bits are similar in construction to those described under rock drilling and tunnelling in Chapters 3 and 10 and diameters exceeding 6 m are available.

Feed force of up to 0.5 kN/mm bit diameter may be required. A formula developed by the Hughes company gives torque required as:

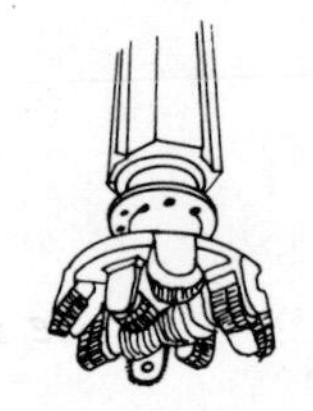

ROLLER BITS FOR ROCK DRILLING

Fig. 5.24 Roller bits for rock drilling

$$T = 5252 \times k \times D^{2.5} \times W^{1.5}$$

where D = bit diameter in inches
T = torque in lb/ft
W = load per inch on the bit in 1000 lb.
k = formation constant, e.g. 4×10^{-5} for granite.

For example a 60 inch (1.5 m) diameter bit would require a torque of 20 000 lb/ft (30 kN.m).

Output

Medium hard rock 0.5 – 1 m/h

Typical data for soft/medium rock drilling are given in Table 5.4.

Table 5.4 Various drilling requirements for effective penetration when circulation drilling with a full hole bit in medium soft rock formations

Hole diameter (mm)	Flushing rate (l.p.m.)	Weight of drill stem (tonnes)	Rotary speed (r.p.m.)
750	6000	20	30–50
1000	8000	25	25–35
1250	10000	38	20–30
1500	12000	40	15–25
1750	16000	47	10–20
2000	20000	55	10–20
2250	24000	61	10–15
2500	28000	68	10–15
3000	32000	80	7–12
4000	36000	100	5–10

References

Atlas Copco Company (1983) *Auger Drilling*, Daventry, Northants.

Bowman, I. (1911) 'Well drilling methods', Geological Survey Water-supply Paper, **257**, Washington, Govt. Printing Office, USA.

BSP International Ltd. (1983) *Earth Drilling Equipment*, Ipswich, England.

B.S. 5930 (1981) *Code of Practice for Site Investigation*.

Calweld, Smith International Inc., (1983) *Tunnelling and Drilling Equipment*, Santa Fe Springs, California, USA.

Craelius Ltd., (1983) *Auger Drilling*, Daventry, England.

Gordon, R.W. (1958) *Water Well Drilling with Cable Tools*. Ruston Bucyrus Co., USA.

Hughes Tool Company (1983) *Williams Construction Diggers*, Dallas, Texas, USA.

Huisman, L. (1972) *Groundwater Recovery*, Macmillan, London.

Junickis, A.R. (1971) *Foundation Engineering*, International Textbook Company, USA.

Tomlinson, M.J. (1978) *Foundation Design and Construction*, Pitman, London.

Winterkorn, H.F. and Fang, H.Y. (1975) *Foundation Engineering Handbook.* Van Nostrand Reinhold Ltd., New York, USA.

CHAPTER SIX

De-watering of ground

Methods of dealing with ground water

In order to carry out construction work below surface levels it is normally necessary for the working area to be reasonably free from standing water. Therefore the ground water flow must either be blocked or carried away. The type of soil, the height of the water table, the depth of the excavation and its shape all influence the choice of method of dealing with water flow in and around the excavation works and a variety of methods is available. For example, the water may be removed by pumping, or it may be isolated from the works by providing a barrier of injected material to fill the soil interstices, or the water in soil voids may be frozen, or compressed air may be used to provide a pressurised chamber to balance the water head in the ground.

It can be seen in Fig. 6.1 that for the majority of granular soils, pumping methods are most appropriate. For medium to coarse gravels

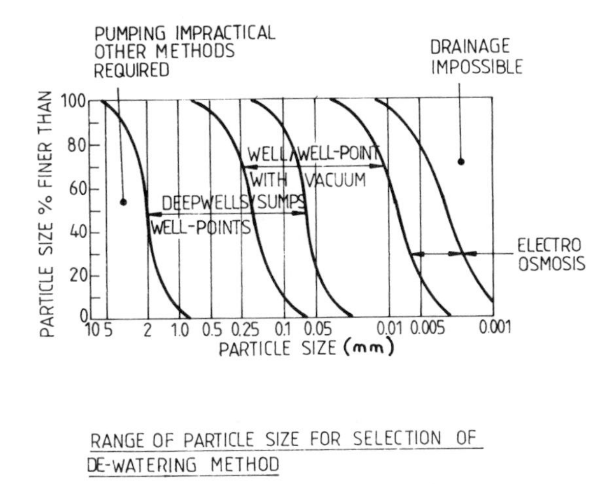

Fig. 6.1 Range of particle size for selection of de-watering method

however, permeability of the material is usually too high for pumping, and cement grouting may be necessary. At the other extreme, clay is an impermeable soil and therefore water seepage would not be a major problem, any surface run-off could be adequately removed from a sump. Silty soils are particularly troublesome and often cannot be effectively de-watered by pumping. Either electro-osmosis, or more usually, ground freezing or grouting should be considered. Freezing and grouting are more adequately covered in Chapters 7 and 9 and compressed air in Chapters 9 and 10.

This chapter deals with pumping methods only.

Theory of de-watering soils

The objective of de-watering is to lower the water table in the vicinity of an excavation to provide a relatively dry and stable working area. Pumping from wells positioned outside the excavation boundary is usually the preferred technique. The type of soil and the position of impermeable strata have a marked effect on the rate of pumping which is possible and theoretical models have been developed to estimate the discharge from a well in different soil configurations. However, to simplify the formula for practical applications, the flow into a well is usually assumed to be either confined or unconfined.

Steady confined flow to a single well

Darcy's Law states that the quantity of water flowing under laminar conditions is proportional to the hydraulic gradient, i.e.

$$q \propto -\frac{ds}{dr} \tag{6.1}$$

Thus for the well shown in Fig. 6.2,

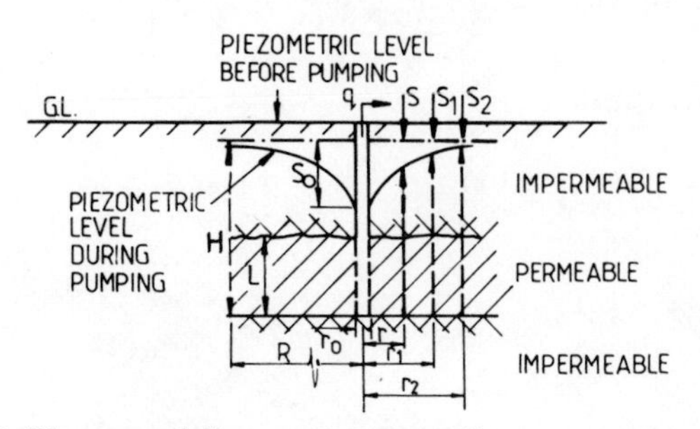

Fig. 6.2 Flow to a fully penetrating well in an artesian aquifer

$$q = -k2\pi rL \frac{ds}{dr} \tag{6.2}$$

where r = radius from centre of well to a given point

L = thickeness of permeable soil

q = theoretical quantity of flow per unit time through a cross-sectional area at radius r

k = constant = coefficient of permeability of the soil, defined as quantity of water that will flow through a unit cross-section of porous material in unit time under a hydraulic gradient of unity.

$-\frac{ds}{dr}$ = hydraulic gradient at radius r.

After integrating, $s = -\frac{q}{2\pi kL} \log_e r + C$

Thus the draw-down at $S_1 = -\frac{q}{2\pi kL} \log_e r_1 + C$

and draw-down at $S_2 = -\frac{q}{2\pi kL} \log_e r_2 + C$

Therefore

$$S_1 - S_2 = \frac{q}{2\pi kL} \log_e \frac{r_2}{r_1} \tag{6.3}$$

The equation is more generally written as:

$$S = \frac{q}{2\pi kL} \log \frac{R}{r}$$

This is Dupuit's formula, where R is the radius at the tangent point between the pumped piezometric level (draw-down curve) and the original water table.

Steady unconfined flow to a single well (*Fig. 6.3*)

Using Darcy's formula

$$q = 2\pi rh \times k \times \frac{dh}{dr} \tag{6.4}$$

rearranging

$$hdh = \frac{q}{2\pi k} \times \frac{dr}{r}$$

and integrating

$$h^2 = \frac{q}{\pi k} \log_e r + C$$

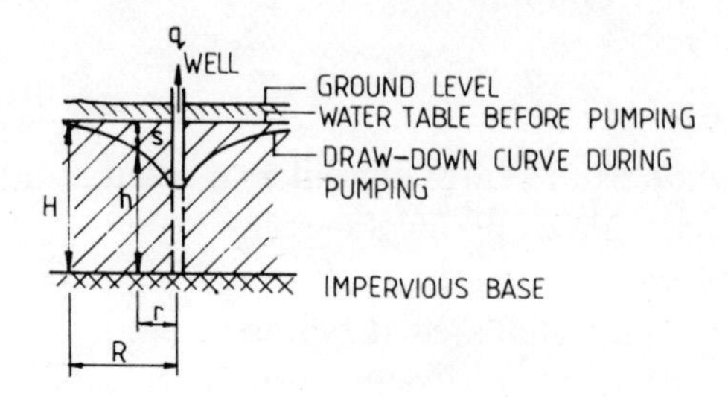

Fig. 6.3 Flow to a fully penetrating well in an unconfined aquifer

This formula is generally written as

$$H^2 - h^2 = \frac{q}{\pi k} \log_e \frac{R}{r} \qquad (6.5)$$

which is the Dupuit formula.

The draw-down at a radius r is $S = H - h$ and therefore, substituting for h into (6.5)

$$S = \frac{q}{\pi k (2H - S)} \log_e \frac{R}{r}$$

For small draw-downs in deep aquifers, S is small compared to H and can be neglected and so

$$S = \frac{q}{2\pi k H} \log_e \frac{R}{r} \qquad (6.6)$$

Thus the difference in draw-down between two such points S_1 and S_2 can be shown by

$$S_1 - S_2 = \frac{q}{2\pi k H} \log_e \frac{r_2}{r_1} \qquad (6.7)$$

This is Thiem's formula.

Partially penetrating wells in deep aquifers (*Fig. 6.4*)

Excavations are often fairly shallow, perhaps 2 – 15 m in depth. Therefore where the impermeable stratum is relatively deep the Dupuit-Forcheimer formula is modified to:

$$q = \frac{\pi k (T^2 - (T - H + h_0)^2)}{\log_e R - \log_e r_0} \times \sqrt{\frac{H}{T}} \times \sqrt[k]{\frac{2T - 1}{T}} \qquad (6.8)$$

Other confined soil configurations

The Dupuit formulae described for a well founded in a completely

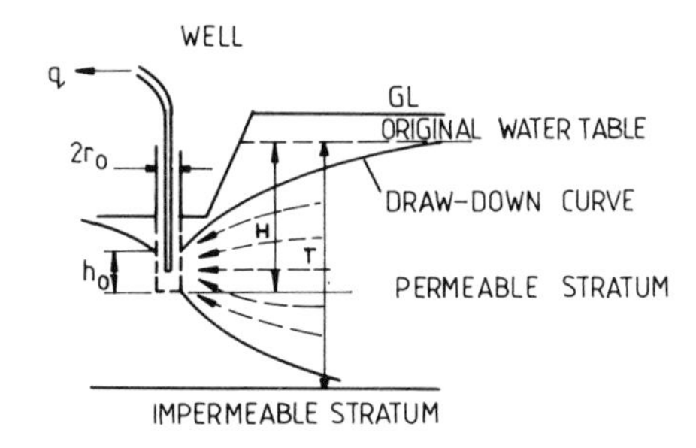

Fig. 6.4 Flow to a well founded above an impermeable stratum (Jureka)

confined or unconfined stratum can be further developed to embrace layered impervious and porous strata. This is beyond the scope of this book and the reader is recommended to follow *Groundwater Recovery* by Huisman for further information.

Determining well discharge

In order to assess accurately the discharge from a well and its associated draw-down curve, for example in water supply calculations or where complex non-uniform soil strata are involved, it is necessary to carry out field measurements on a trial borehole and take measurements from stand pipes (Fig. 6.5). The method requires pumping until steady state conditions are obtained, which may take many weeks.

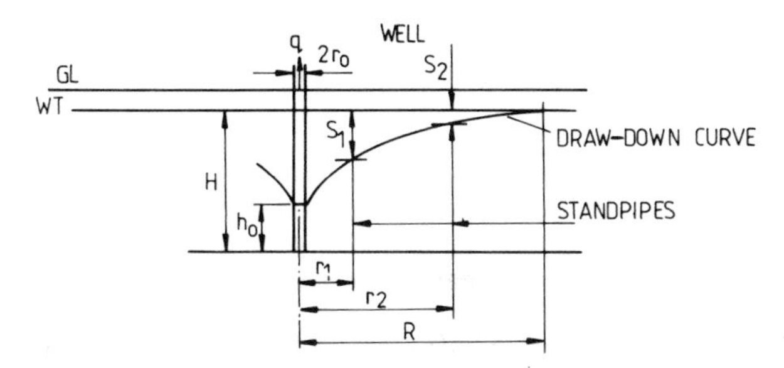

Fig. 6.5 Determining well discharge from pumping trials

However, for estimating purposes the above Dupuit formulae can be used to give the approximate pumping rate necessary to produce an assumed draw-down level. Thus, for given values of k, H, h_0, R and r_0, the pumping rate from a well may be calculated.

Approximate coefficients of permeability based on field experience are given in Table 6.1. Typical values for R are shown in Fig. 6.6.

Table 6.1 Soils and their approximate values of permeability

Soil	Grain size (mm)	Coefficient of permeability (mm/s)
Medium gravel	4 - 7	>30
Fine gravel	2 - 4	30 - 60
Uniform sand	0.11	2 - 0.05
Graded sand	0.1 - 0.3	$1 - 10^{-2}$
Silty sand	Varies	$10^{-1} - 10^{-2}$
Clayey sand	Varies	$10^{-2} - 10^{-3}$
Silt	0.05 - 0.002	$10^{-3} - 10^{-5}$
Clay	0.002	$10^{-5} - 10^{-6}$

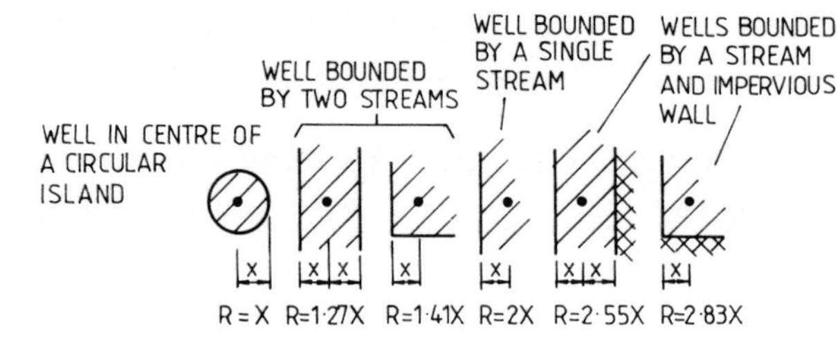

Fig. 6.6 Approximate values of maximum radii of draw-down curves (Huisman)

Well losses

The Dupuit formulae assume that the water flows in horizontal planes. At large radii from the well, Fig. 6.7 shows this assumption to be reliable. Near the well, however, the flow is curvi-linear and the free surface and Dupuit's assumed curve do not coincide. An estimate of difference in height (m) between these two curves at the well periphery (Fig. 6.8(a)) is given by Ehrenberger as:

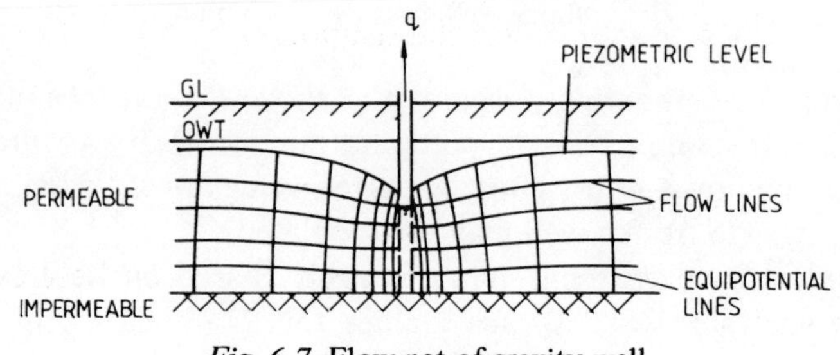

Fig. 6.7 Flow net of gravity well

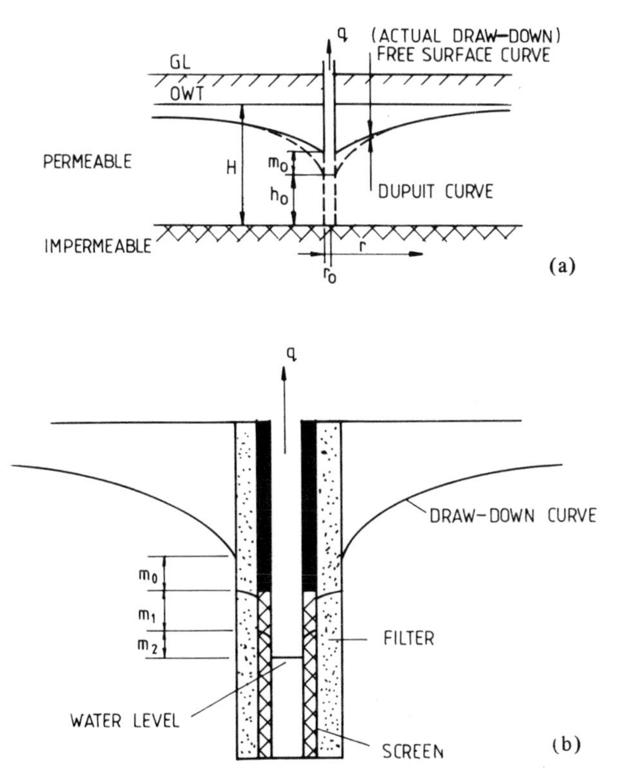

Fig. 6.8 Head losses at the entrance to a well

$$m_o \doteq \frac{H}{2}\left(1 - \frac{h_o}{H}\right)^2 \tag{6.9}$$

Also, in practice, additional losses are caused by entrance friction at the screen (m_2) and through the filter (m_1) (Fig. 6.8(b)). Therefore the actual rate of discharge from the well is likely to be considerably less than the theoretical.

Sichardt has suggested that the actual hydraulic gradient (i) at the entrance to the well is related to the coefficient of permeability (k) by

$$i = \frac{1}{15\sqrt{k}} \text{ for the temporary wells}$$

and

$$i = \frac{1}{30\sqrt{k}} \text{ for permanent, long standing wells.}$$

From Darcy's law, the entrance velocity $v = ki$. Therefore for a temporary well

$$v = \frac{\sqrt{k}}{15}$$

Thus for a well with radius r_o and screen depth h_w the actual discharge obtained would be

$$q_w = 2\pi r_o h_w \frac{\sqrt{k}}{15} \tag{6.10}$$

Consequently, to obtain the desired draw-down level, either a greater well radius than the theoretical, or several wells are required.

Discharge from an unconfined well system

In de-watering construction works it is often more economical to install a group of small wells rather than a single large well. Interference between two or more wells occurs when the cones of draw-down overlap. The draw-down at any point on the composite cone of depression (i.e. the combined draw-down curve of several wells) is then equal to the sum of the draw-downs at that point for each well, assuming it to be pumped separately as shown in Fig. 6.9.

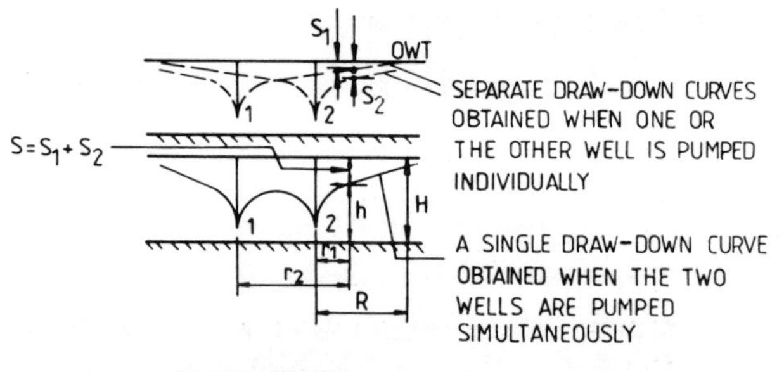

Fig. 6.9 Interference between adjacent draw-down curves

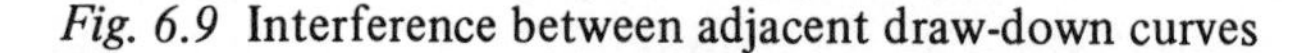

Thus at any radius r from a well in a group of wells, the draw-down at r is

$$S = \sum_{i=1}^{i=n} S_i$$

From equation (6.5),

$$H^2 - h^2 = \sum_{i=1}^{i=n} (H^2 - h_i^2) = \sum_{i=1}^{i=n} \frac{q_i}{\pi k} \log_e \frac{R}{r_i}$$

If $q_1 = q_2 \ldots = q_n = q$ i.e. discharge is the same for all wells where q is the discharge from a single well, then

$$H^2 - h^2 = \frac{q}{\pi k}(n \log_e R - (\log_e r_1 + \log_e r_2 \ldots \log_e r_n))$$

Thus

$$H^2 - h^2 = \frac{q}{\pi k}(n \log_e R - \log_e r_1 \times r_2 \times \ldots r_n) \qquad (6.11)$$

Wells arranged in a circle

$$r_1 = r_2 = \ldots r_n$$

Therefore

$$H^2 - h^2 = \frac{nq}{\pi k}(\log_e R - \log_e r) \qquad (6.12)$$

If the wells are arranged around the excavation (see Fig. 6.10) the required radius of the circle $x \triangleq \sqrt{\dfrac{b \times l}{\pi}}$

where b = breadth of excavation
l = length of excavation.

Example (*Fig. 6.10*)

Wells are founded in a circular pattern around an excavation 30 m long and 25 m wide. The height of the water table above an impermeable stratum is 9 m. The depth of the excavation is 5 m and the water table lies 1 m below ground level. The coefficient of permeability of the soil $k = 0.002$ m/s. Determine the required number of wells.

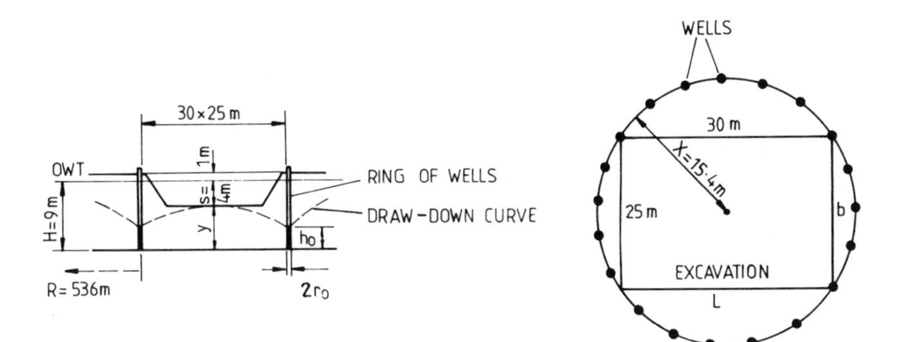

Fig. 6.10 Design example

Solution

From equation (6.11) and when $r = x$

$$H^2 - h^2 = \frac{nq}{\pi k}(\log_e R - \log_e x) \qquad \text{(A)}$$

(i) $x = \sqrt{\dfrac{b \times l}{\pi}} = \sqrt{\dfrac{25 \times 36}{\pi}} = 15.4 \therefore \log_e x = 2.73.$

(ii) Values of R can be obtained from Fig. 6.6, but in open country, Sichardt has shown that for an excavation

$$R \triangleq 3000 \times S \times \sqrt{k}$$

where k is in m/s.

Thus $R = 3000 \times 4 \times \sqrt{0.002} = 536$ m. as shown in Fig. 6.10

$$\log_e R = 6.28$$
$$h = H - S = 9 - 4 = 5 \text{ m.}$$

(iii) Therefore substituting in (A)

$$nq = \frac{\pi k\,(H^2 - h^2)}{\log_e R - \log_e x} = \frac{\pi \times 0.002\,(9^2 - 5^2)}{6.28 - 2.73} = 0.1 \text{ m}^3/\text{s.}$$

(iv) If the effective length of screening on a well of diameter r_o is h_o

$$\text{then the capacity of the well } q = 2\pi r_o h_o \times \frac{\sqrt{k}}{15}$$

Assuming the effective length of screening is approximately $0.4H$ and r_o is 100 mm then, substituting values into the equation

$$q = 2 \times \pi \times 0.1 \times 0.4 \times 9 \times \frac{\sqrt{0.002}}{15} = 0.0067 \text{ m}^3/\text{s or } 6.7\ l/\text{s}$$

Therefore the number of wells required is

$$n = \frac{0.1}{0.0067} = \underline{15 \text{ wells}} \quad \text{plus approx. 5 wells for reserve} = 20 \text{ wells}$$

(v) Well diameter: Sichardt recommends that the minimum spacing of wells should be about $10r_o\pi \triangleq 32r_o$.

Thus with $x = 15.4$ m (Fig. 6.10),

$$\text{spacing of wells} = \frac{2\pi \times 15.4}{20} = 4.8 \text{ m} > 32 \times 0.1 > 3.2.$$

Thus the selected well diameter of 200 m is suitable. A slightly smaller diameter would also be acceptable, although the costs of producing additional boreholes and the extra pumps must be taken into account.

Pumping from open excavations

Open sumps

The system of pumping from an open sump is popular because the costs of installation and maintenance of the equipment are relatively low compared to those for wells, and because the system is applicable to most soils. Unfortunately, the method draws water into the excavation, because the sump is usually located inside the excavation itself. As a consequence, the tendency is to wash out the banks when working in fine soils. Nevertheless with due care, coupled with support for the sides of the excavation, pumping may be adequate in providing a reasonably water-free working area. Dense and well-graded granular soils, hard fissured rocks and surface run-off from clays are favourable to open pumping, but problems with slope stability are likely to occur in loose granular soils, soft granular silts and soft rock.

Open excavation in an unconfined soil founded on an impermeable stratum (*Fig. 6.11*)

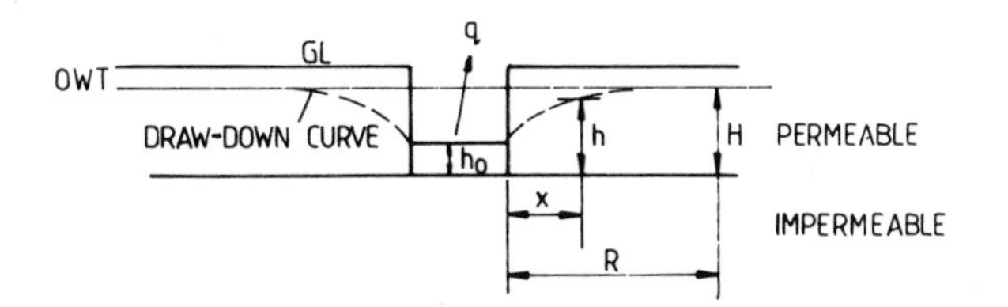

Fig. 6.11 Open excavation in an unconfined soil founded on an impermeable stratum

The theory of pumping from sumps can be developed along similar lines to wells on the assumption of steady flow from a confined or unconfined aquifer.

According to Darcy, the rate of flow of water through a unit length of sump wall is

$$q = kh\frac{dh}{dx} \quad \text{or} \quad hdh = \frac{q}{k}dx$$

and integrating

$$\frac{h^2}{2} = \frac{qx}{k} + C$$

when $\qquad x = \text{o},\ h = h_\text{o} \therefore C = \frac{h_\text{o}^2}{2}$

when $x = R \quad h = H$

and so

$$\frac{H^2}{2} = \frac{qR}{k} + \frac{h_o^2}{2}$$

Therefore
$$q = \frac{k}{2R}(H^2 - h_o^2) \tag{6.13}$$

Jureka suggests that length of draw-down curve from the sump face to the tangent to the water table be represented by the equation

$$R = 2H\sqrt{kH} \quad \text{where } k \text{ is in m per day.}$$

Open excavation in an unconfined soil above an impermeable stratum (*Fig. 6.12*)

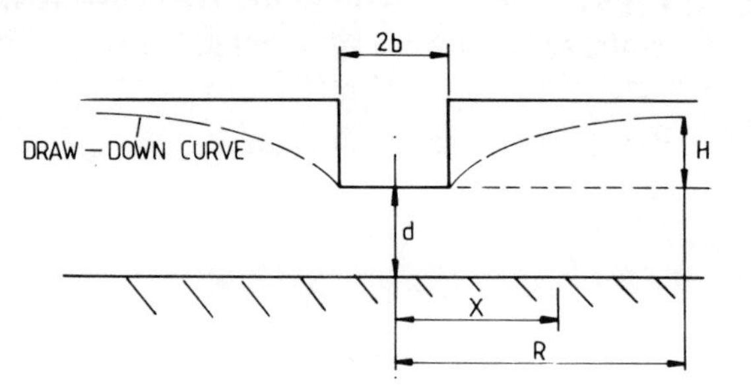

Fig. 6.12 Open excavation in an unconfined soil above an impermeable stratum

The simple situation above is rarely found in practice and it is more usual for an excavation to be located above an impermeable stratum. Jureka recommends that eqn. (6.13) be adjusted to take into account seepage through the base as follows:

$$q = \frac{kH}{2}\left[\frac{H}{R} + \frac{\pi}{\log_e \dfrac{d}{\pi b} + \dfrac{\pi R}{2d}}\right] \tag{6.14}$$

Open excavation in a confined soil above an impermeable stratum (*Fig. 6.13*)

$$q = k\left[\frac{(2s - m)}{R}m + \frac{\pi s}{\log_e \dfrac{d}{\pi b} + \dfrac{\pi R}{2d}}\right] \tag{6.15}$$

As with wells, the actual rate of flow (q_s) into the sump will be significantly lower than the theoretical value q.

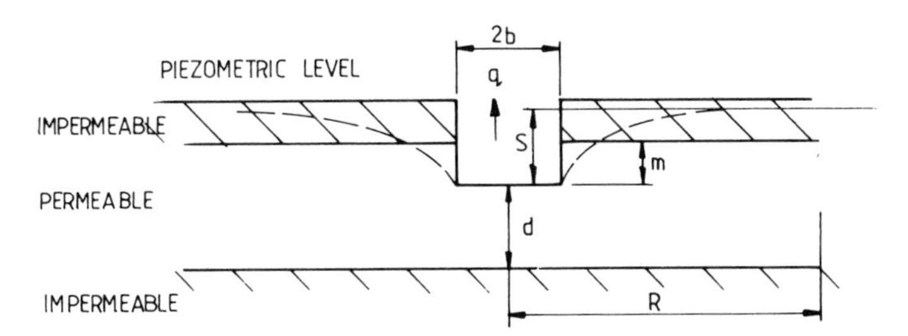

Fig. 6.13 Open excavation in a confined soil above an impermeable stratum

Closed excavations founded above an impermeable stratum

The flow of water entering an excavation can be reduced by enclosing the sides, for example with sheet piles. Ideally the piles should be driven down to an impermeable stratum to seal the excavation completely, as shown in Fig. 6.14. Where this is impracticable, seepage through the base must be expected (Fig. 6.15) and the rate of flow per

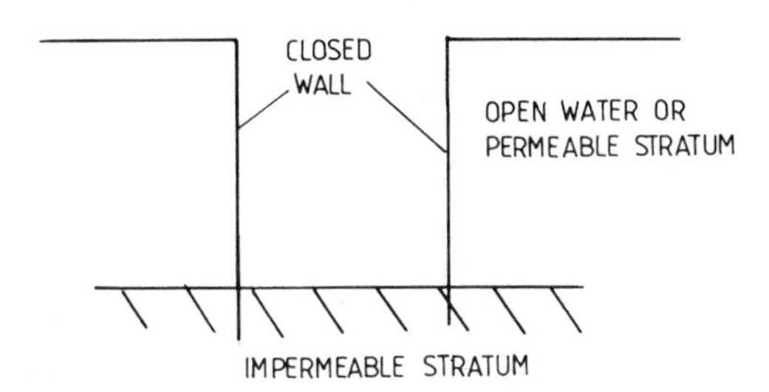

Fig. 6.14 Closed excavation founded on an impermeable stratum

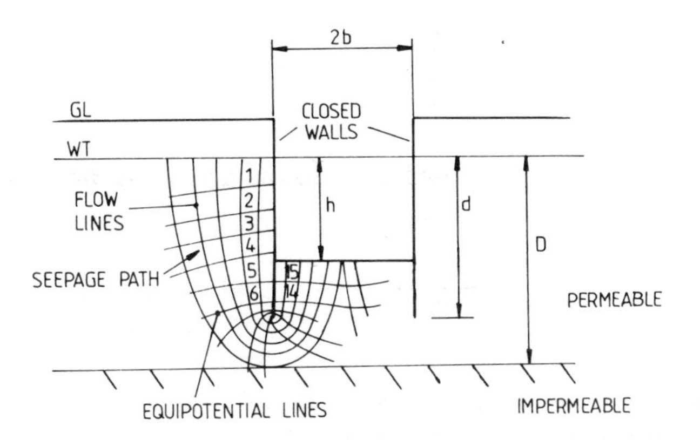

Fig. 6.15 Closed excavation founded above an impermeable stratum

unit length is given by:

$$q \triangleq b\kappa i \qquad (6.16)$$

where i is the hydraulic gradient as determined from flow net analysis.

However this formula is only accurate when d is much less than D. If this relationship is altered by driving the piles deeper, then the rate of flow in the formula should be reduced by deducting the following values from q.

$\frac{d}{D}$	0.1	0.2	0.3	0.4	0.5	0.6	0.7	0.8	0.9	1.0
% deducted from q	5	10	15	20	25	30	40	50	65	100

'Piping' or 'boiling' of the soil inside the excavation

Piping will occur when the upward force of the water issuing is greater than the weight of the particles.

The upward force of the water $= \gamma_w i$

The weight of the particles in water $= \gamma_w \left(\frac{P_p - 1}{1 + e}\right)$

where γ_w = unit weight of water
P_p = specific gravity of the particles
e = voids ratio.

Therefore piping occurs when $i > \frac{P_p - 1}{1 + e}$

Example (*see Fig. 6.15*)

For sand, $P_p \triangleq 2.8$ and $e = 0.8$.

$$\frac{P_p - 1}{1 + e} = \frac{2.8 - 1}{1.8} = 1.$$

Thus if $h = 21$ m, $n = 15$, the minimum number of cells in a path, each of a side length $a = 1$, is as follows:

$$i = \frac{h}{n \times a} = \frac{21}{15 \times 1} = 1.4 > 1 \text{ and piping will occur.}$$

In practice a factor of safety significantly in excess of 1 is desirable, because the exit velocity in highly permeable soils may also cause movement of the soil particles.

Piping usually does not occur in clays and silty clays, because seepage either does not take place or is quite small. It is also unlikely to occur in gravels because the high permeability tends to allow extensive draw-down and thereby produces a naturally long seepage path. Piping occurs most often in loose fine sands which have relatively high permeabilities with concentrated draw-down curves, and whose small grains are also susceptible to forces caused by the flow of water. In these conditions excavation work can be most hazardous and, rather than the use of sumps, external de-watering of the soil may be required, together with cut-off walls. Unless these methods are effective, soil particles will be taken up by the flow into the excavation and the surrounding banks may suffer subsidence.

Sump design

A sump is positioned in the deepest part of the excavation and preferably away from the main works. A small ditch cut around the base of the excavation falling towards the sump will help to keep the area reasonably clear of standing water. If the work is to continue for a period then it often pays to use porous pipes in a gravel fill in the ditch.

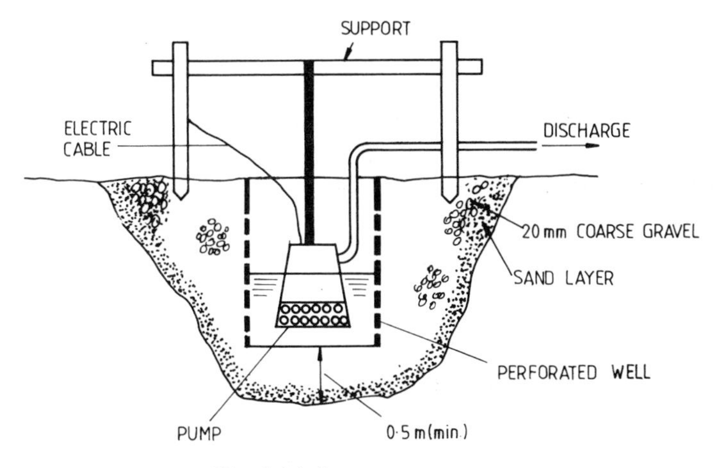

Fig. 6.16 Sump arrangement

The sump (Fig. 6.16) base should be approximately 1 m below the bottom of the excavation, with the walls protected with timber, loose trench sheets, a perforated oil drum, etc. The bottom of the sump can be stabilised with a 0.5 m layer of graded granular fill, e.g. sand to coarse gravel. If this aspect is overlooked, fines in the soil may be washed through and damage the pump.

Well pointing

In permeable strata where soil stability might be endangered by using sumps, the single stage well pointing system is a preferred method for shallow excavations not exceeding 6 – 7 m deep. For greater depths, multi-stage well pointing or a single, large well should be considered. The objective of the well point system is to produce a cone of depression in the water table so that excavation can take place in relatively dry conditions. The system consists of a number of individual well points, each comprising a jetting/riser pipe, 40 – 50 mm diameter, drilled with a ring of inlet ports at the bottom (Fig. 6.17). A strainer about 1 m long is placed over the tube to cover the ports. The riser is connected at surface level to a header pipe, about 150 mm diameter, which in turn is connected to a suction pump.

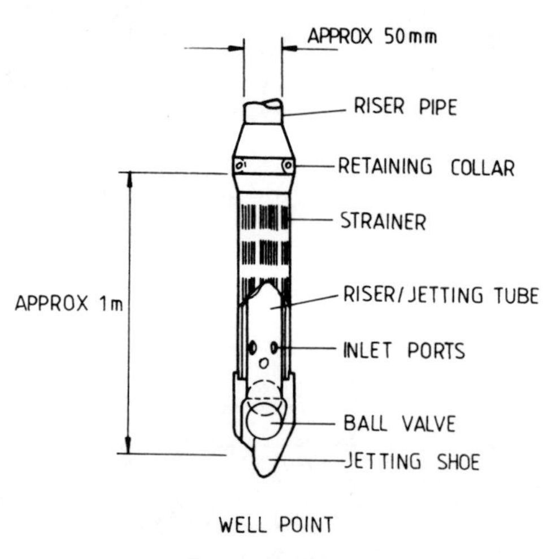

Fig. 6.17 Well point

Installation

The well point is fitted with a rubber ball valve placed inside the jetting shoe (see Fig. 6.17) and during installation the top end of the riser/ jetting pipe is connected to a jetting hose and water under pressure is forced through the well point. Usually maximum pressure of 1 N/mm^2 (10 bar or 150 p.s.i.) is sufficient in most types of soil, with a pump capable of delivery of 100 – 120 *l*/min.

The procedure simply requires that an operator place the point in the desired position (Fig. 6.18). The water pressure is turned on and the washing or jetting action causes the pipe to penetrate the soil. The water is turned off when the correct depth is located. The operation is then repeated until all well points are in position.

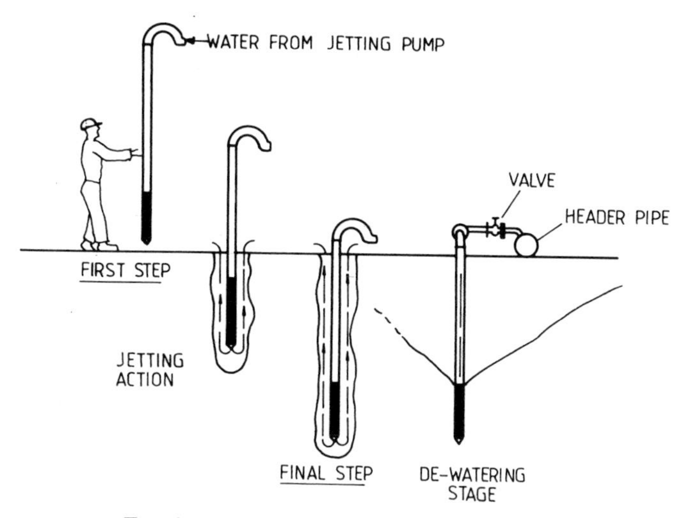

Fig. 6.18 Installing well points and riser pipes

Well pointing arrangements

Progressive systems (*Fig. 6.19*)

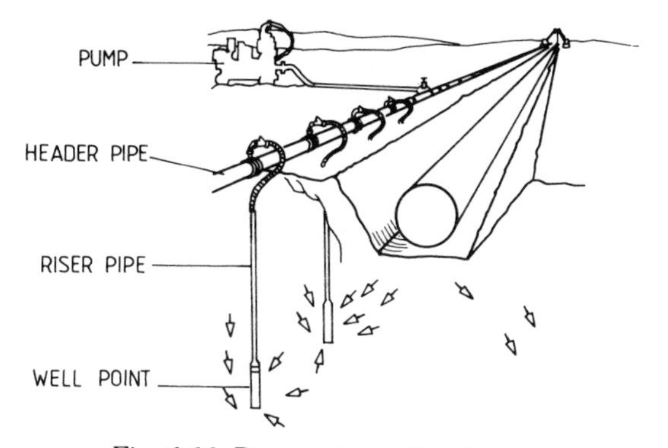

Fig. 6.19 Progressive well point system

For trench work a header pipe is placed along the line of the proposed trench outside the track of the excavator. Well points are progressively installed ahead of the excavation. By inserting isolating valves in the header pipe and repositioning the suction pump, the excavation may be progressively backfilled. The trenches can be timbered, sheet piled or battered in the usual manner to improve stability as demanded.

The number of well points needed can be theoretically calculated as described on page 103, but in practice 750 mm centres are typical in

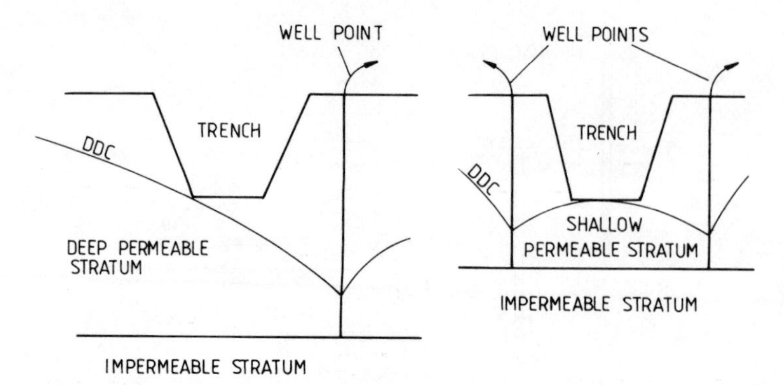

Fig. 6.20 Effects of depth of foundation and impermeable stratum on choice of de-watering arrangement

loose gravel or coarse sand, while the spacing may be increased in fine running sand.

In deep permeable soil a single row of well points may be adequate, but if the trench is founded close above an impermeable stratum then well points both sides of the trench may be necessary to obtain draw-down below the bottom of the excavation (Fig. 6.20). In these latter conditions it may be impossible to secure a completely dry base, as some water will inevitably pass between points.

An installation gang of approximately 4 men is usually sufficient to keep ahead of a trenching gang.

Ring system (*Fig. 6.21*)

In this system the well points remain in position over the full construction period. The header pipe is placed at surface level around the perimeter of the excavation to form a closed ring. Well points are then sunk into position. Where more than a single pump is required to cope with the rate of flow, each pump should be located to house an equal number of points either side.

Staged ring system

In theory, atmospheric pressure limits suction to about 10.35 m but because of leakages in pipe joints etc. the maximum depth for a single stage system is about 6 – 7 m. Well pointing of deeper excavations, however, can be achieved with additional rings, as shown in Fig. 6.21. Each stage lifts within the limiting depth, and water is pumped to surface level with a separate pump located at the stage level. The method requires a wide area to accommodate the stages and when the

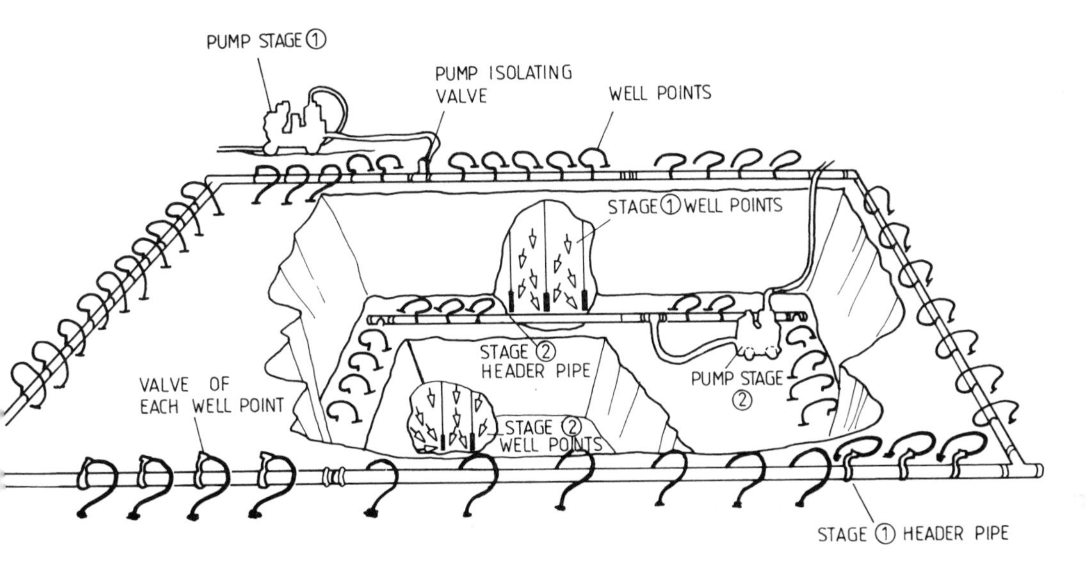

Fig. 6.21 Ring well pointing system (also shows staged rings)

costs of the additional excavation work are taken into account, the alternative method of using deep wells might be a cheaper alternative.

General points in well point operation

(i) Install one or more 'dummy' wells (i.e. open-ended pipes) to monitor the draw-down level.

(ii) It is undesirable to commence excavation work before the water table has reached the design depth and remained steady for a period.

(iii) Pumping should be continued until permanent structures are completely above standing water level. Hence standby pumps are required.

(iv) If output from the pump is small, but the vacuum gauge has a high reading, then the well points are probably blocked. This problem can sometimes be overcome by opening the header pipe to atmospheric pressure and allowing the water to flow back to ground.

(v) If vacuum is lost, with an accompanying rise in the water table, then air leaks into the circuit are present. The source of leakage can sometimes be found by isolating the individual valves, one at a time on each of the riser pipes.

(vi) Initially the discharge may be discoloured for a few minutes and then clear. If the discoloration persists, then fines are being drawn into the flow, which may cause damage to nearby structures.

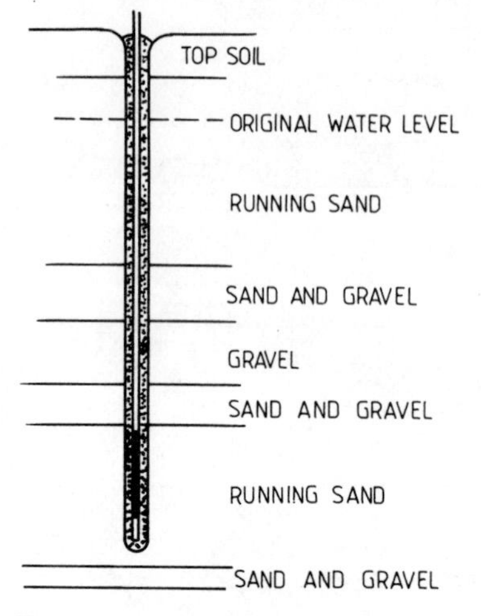

Fig. 6.22 Sanding in well points

(vii) Where permeable soil is interspersed with clay layers, the water table will be lowered, but also water may flow along the impermeable strata and into the excavation. In these conditions the well points should be 'sanded in' as shown in Fig. 6.22. By enlarging the well diameter, a surround of filter material will conduct the flow more efficiently.
'Sanding in' should also be employed in fine silts, whereby the filter medium may assist in preventing the fines passing through the strainer of the well points.

Output data

The maximum capacity of a single well point is approximately 1.0 – 1.51 l/s, and to try to achieve this output the spacing is altered to suit the permeability of the soil, for example:

Clay and other impermeable soil	– cannot be de-watered with well points
Silty sands of low permeability	– 1.5 m centres
Fine to coarse sands	– 0.85 – 1.0 m centres
Sandy gravel	– 0.25 – 0.85 m centres
Coarse gravel	– 0.3 m centres
Very porous ground	– well pointing is unsuitable

Jureka in Germany gives typical field results with a well pointing system, as shown in Table 6.2.

Table 6.2 Well pointing field results in various soils

Surface area (m^2)	Number of points	Draw-down depth (m)	Soil permeability (mm/s)	Total discharge (l.p.s.) Start	Finish	Duration of project (days)
1500	50	2	1.0×10^{-1}	15	10	100
1500	100	4	3.0×10^{-2}	20	12	100
2000	75	3	2.5×10^{-1}	13	10	200
2000	100	15	5.0×10^{-1}	23	18	250
2000	300	13	1.0×10^{-3}	7	4	150
2500	175	1	1.0×10^{-2}	4	3	350
3000	100	3	2.5×10^{-1}	20	10	275

Deep wells (150 - 300 mm diameter)

In situations where staging of well point systems is required (see Fig. 6.21) or where the soil permeability is too high for well pointing practicability, deep wells should be considered (Fig. 6.23), the main

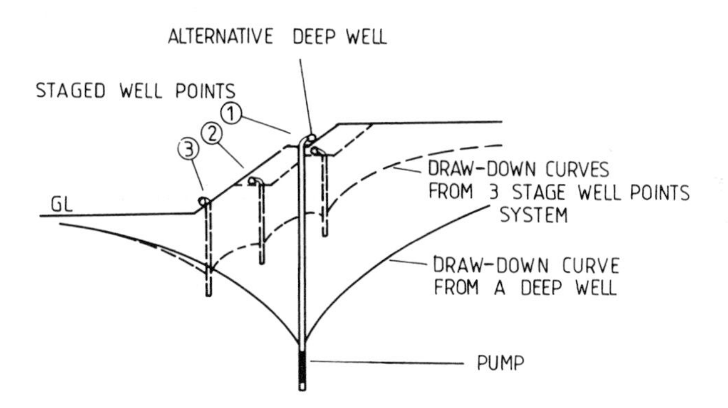

Fig. 6.23 Comparison of well points and deep wells

difference being that a submersible pump is located at the bottom of the well, thereby avoiding the limitations imposed by suction. An appropriate size of pump can be chosen to deal with the rate of flow and the well diameter formed accordingly. Such wells are used for depths greater than about 7.5 m and have been successfully operated at depths greater than 100 m.

Well construction (*Fig. 6.24*)

The well is usually bored with one of the rotary boring methods described

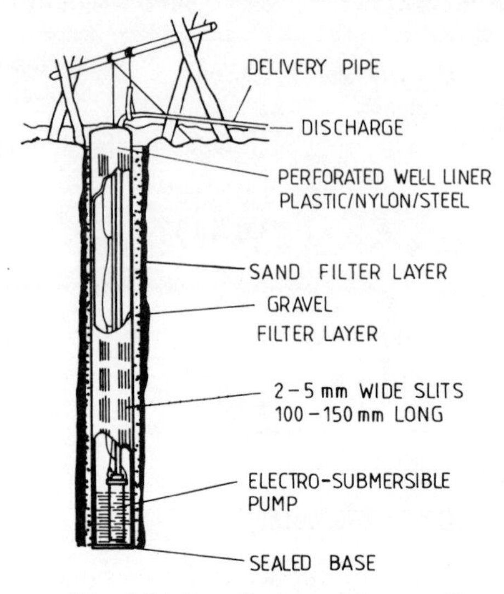

Fig. 6.24 Installation of deep wells

in Chapter 5 and a temporary outer casing is driven to give stability where demanded by the conditions. When the borehole has reached the required depth a perforated well liner is placed into position and plugged at the well bottom to produce a reasonably good seal and stable base. Layers of filter material are placed around the casing to keep out fines.

For the best results the filter material should be cylindrically layered in thicknesses of about 100 mm. The layer near the screen should have grain sizes of 5 - 20 mm diameter, surrounded by a layer with 0.5 - 2 mm grains. The screen itself should be made in plastic, wood, or mild steel sheeting with the perforations placed horizontally or vertically.

To minimise clogging the slits should be as wide as possible; 2 - 5 mm is usual, depending upon the size of the surrounding filter material or soil. (A filter medium is not required in a coarse gravel stratum.) Finally, a submersible pump is installed and the discharge pipe led into a nearby stream.

Output data for shallow tube wells

Typical field results from shallow tube wells are given in Table 6.3.

Table 6.3 Shallow tube wells with results in various soils (Jureka).

Surface area (m^2)	Draw down depth (m)	Number of wells	Soil Permeability (mm/s)	Total discharge (l.p.s.) Start	Finish	Duration of project (days)
250	3.5	10	2.0×10^{-1}	10	5	75
300	3.5	8	2.5×10^{-1}	20	10	180
400	3.5	15	1.5	40	30	200
600	4.0	15	2.5	40	25	180
1000	6.0	50	2.5	120	60	175
2000	2.0	50	5×10^{-1}	30	20	350

Electro-osmosis and vacuum wells

Soils such as silts and silty-clays are virtually impossible to drain with normal pumping methods, because the capillary forces acting on the pore water prevent free flow under gravity. The osmosis technique may therefore be considered as a possible alternative.

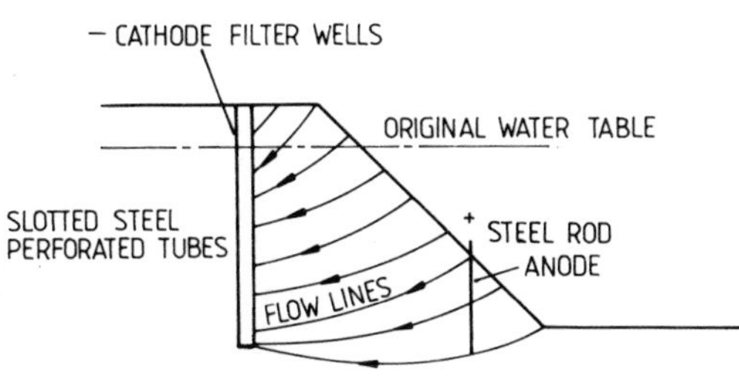

Fig. 6.25 Principles of electro-osmosis de-watering

This method, shown in Fig. 6.25, consists of steel rod anodes and filter wells as cathodes. The positively charged water thus flows towards the filter wells. Unfortunately the system has not been widely used in the U.K. and comprehensive cost data from various soils is not available.

The method is an alternative which may be considered along with vacuum wells or well points (Fig. 6.26), grouting and ground freezing. So-called 'vacuum wells' involve the top part of the riser tube being surrounded with clay to form a seal. The 'vacuum' accelerates the pore water movement.

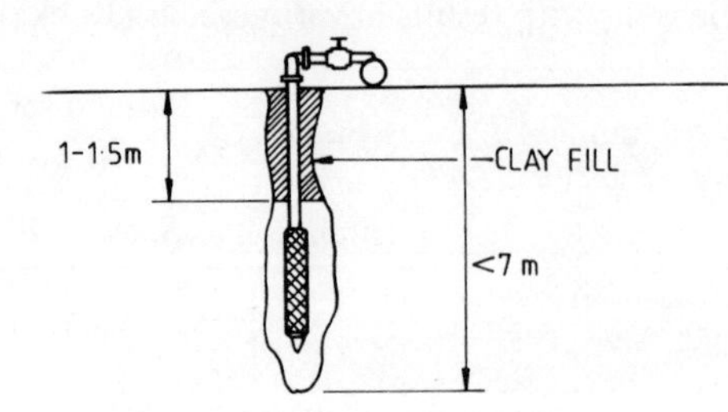

Fig. 6.26 Vacuum well pointing

Recharge water

A lowering of the water table reduces buoyancy of the soil, which may cause settlement of weak, compressible strata. Where settlement must be kept to a minimum, recharge water should be provided. A simple recharge trench (Fig. 6.27) is commonly used in construction works, but reversed well points or recharge wells are alternative methods. However, care is required when choosing the appropriate method. For example, in Fig. 6.28 the recharge water would not effectively

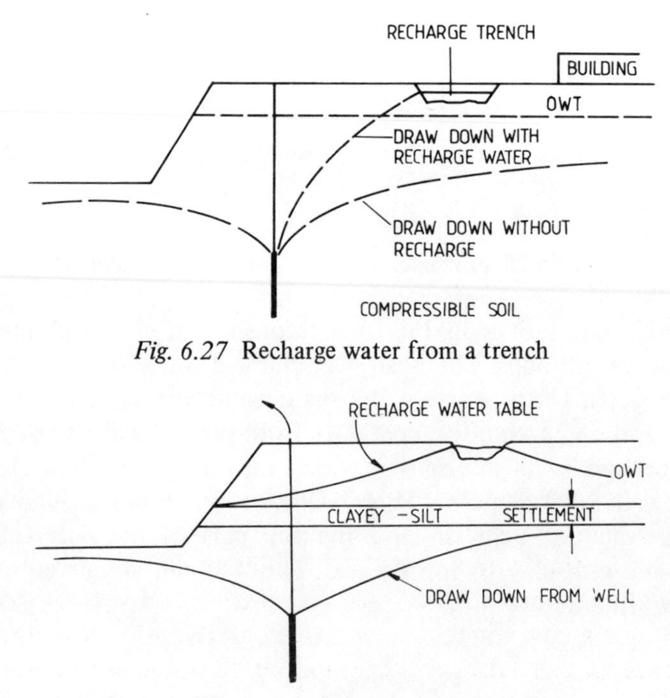

Fig. 6.27 Recharge water from a trench

Fig. 6.28 Example of inappropriate recharging

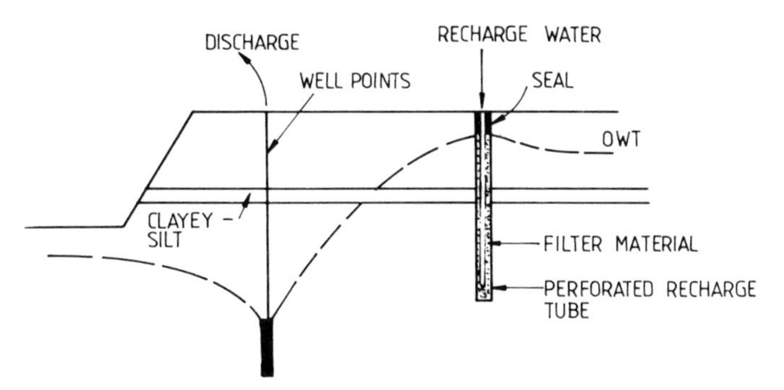

Fig. 6.29 Recharge water from wells

penetrate the compressible layer of clay-silt and might actually increase the load on this layer. Figure 6.29 illustrates the application of a recharge well to try to avoid this effect. The required capacity of a recharge well or trench can be calculated on similar principles to those described earlier for wells and sumps, but with water flowing away from rather than to the well.

Pumping equipment

Lift and force pump (*Fig. 6.30*)

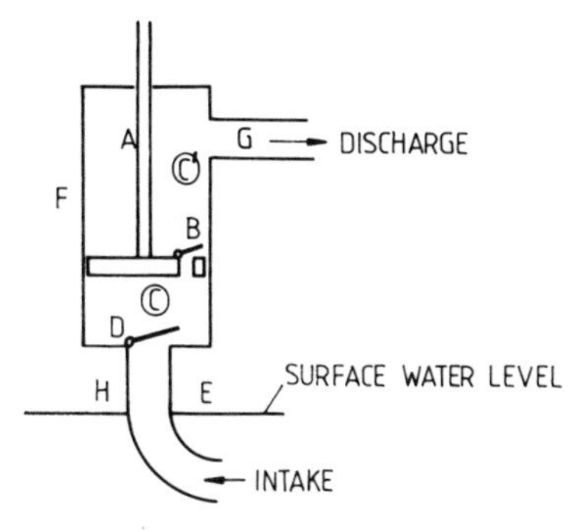

Fig. 6.30 Principles of the lift and force pump

The lift pump depends upon atmospheric pressure for its working principle and comprises a cylinder (F) fitted with a piston (A). A valve is located at the entrance of the cylinder and a valve (B) on the piston. During the upward stroke valve (B) is closed and (D) is opened, thus from Boyle's Law where $p \times v$ = constant, as v is increased the pressure of the air in (C) falls. The pressure of the atmosphere acting on the water surface (E) causes water to flow up the pipe (H) and into (C). On the downward stroke (D) closes and (B) opens and the water is released into (C′). Repetition of the process discharges the water through (G) on the upward stroke to produce a force pump.

In practice, lift and force pumps are operated well within the maximum theoretical atmospheric pressure (taken to be 10.35 m of water) because of leakages in joints etc., and 6 – 7 m is the more normal limiting suction lift. Such pumps are also designed to deliver against a similar head, to give a combined suction and delivery head of about 14 m.

A pump of given power rating produces a rate of discharge which varies with the head as shown in Fig. 6.31.

Such pumps are commonly used in shallow excavations and manufactured with a diesel engine as shown in Fig. 6.32. They are extremely robust and most suited to the variable conditions typical of construction work. 75 mm or 100 mm diameter pipes are usual, fitted with a strainer on the suction end. Such pumps are usually designed (i) to pump on 'snore' i.e. air and water mixed during low flows and (ii) to pick up suction without priming.

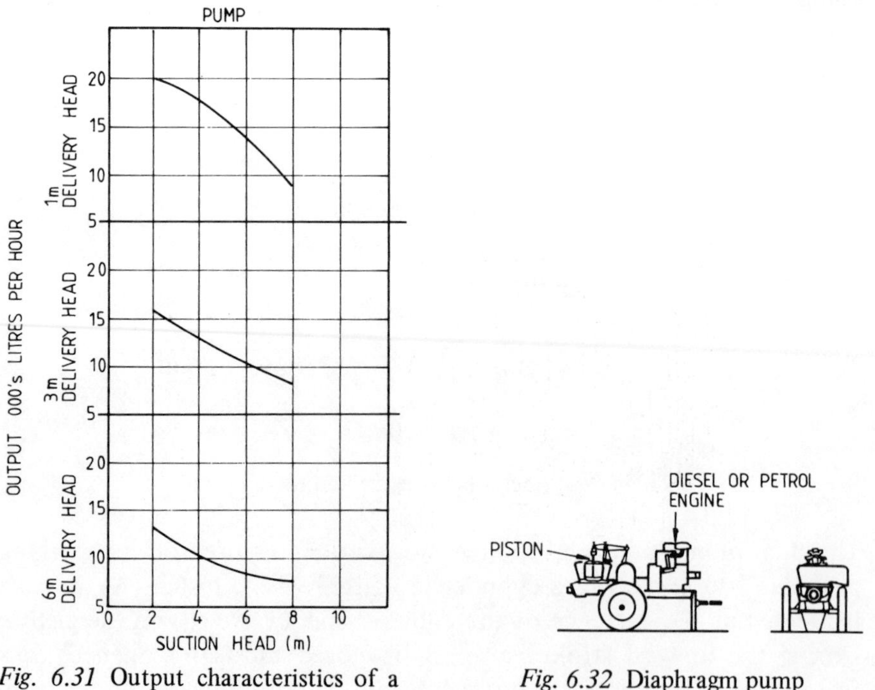

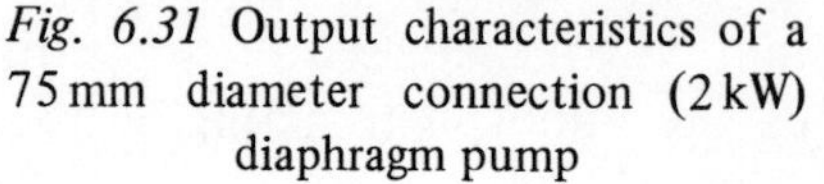
Fig. 6.31 Output characteristics of a 75 mm diameter connection (2 kW) diaphragm pump

Fig. 6.32 Diaphragm pump

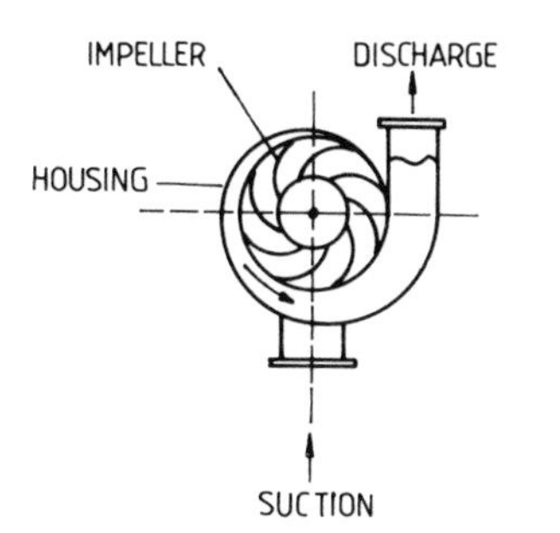

Fig. 6.33 Principles of the centrifugal pump

Centrifugal pump

Water may be raised by the centrifugal rotation of a vane wheel, operating with a reversed turbine action (Fig. 6.33). The pump consists of an impeller surrounded by a spiral casing. If the height of the pump above the water surface is located within the lift permitted by atmospheric pressure, water will be drawn inside the casing as the impeller is rotated, imparting additional kinetic energy to it. However, because the casing is spiral in shape, the velocity of the water decreases with increasing area of flow, and on reaching the delivery pipe the velocity will be relatively small, whereas the pressure will be large. In this manner, water can be pumped against extremely high delivery heads.

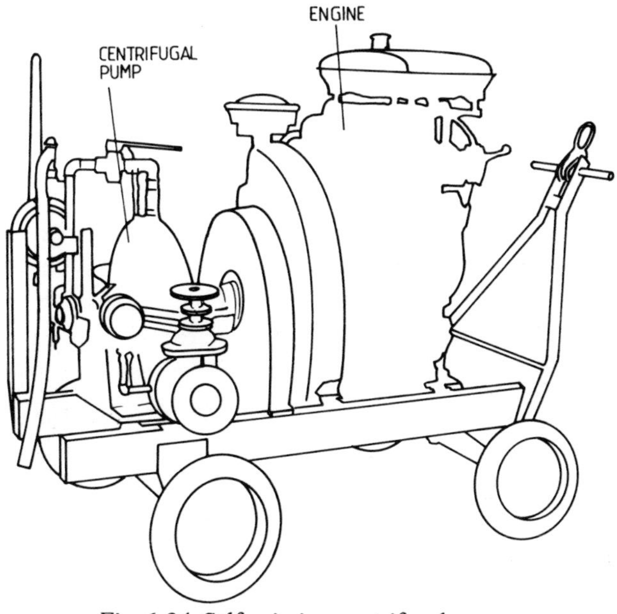

Fig. 6.34 Self priming centrifugal pump

Self priming pump (*Fig. 6.34*)

Centrifugal pumps are commonly used with well pointing, and in sumps they should be of the self-priming type. A 150 mm diameter pump is usually suitable for average well pointing requirements with characteristics typical of those shown in Fig. 6.35. Where a

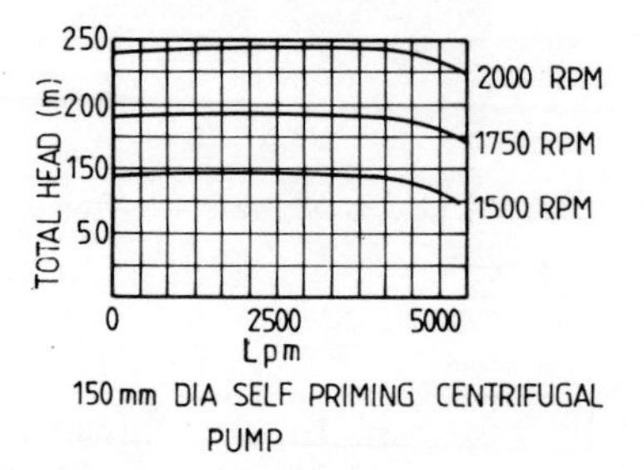

Fig. 6.35 Output characteristics of 150 mm diameter connection self priming centrifugal pump

long suction line and many well points are involved, leaks in the connections and air in the water and ground might require a combined centrifugal water pump and vacuum pump to handle the large quantities of air (see Fig. 5.23, and compressed air, Chapter 2) and a float chamber to separate the air from the water. Such an arrangement is virtually essential in vacuum well pointing.

Submersible pumps (*Fig. 6.36*)

The electric submersible centrifugal pump has gained wide acceptance because of its portability and its ability to deal with water in both sumps and deep wells. Priming is not required and models are now manufactured to suit wells of as little as 250 mm diameter. Larger units are available up to 100 kW power. The characteristics of a typical pump suitable for a deep well are shown in Fig. 6.37. The main disadvantage with this type of pump is the wear caused to the impeller blades by granular particles carried in the water, which results in loss of capacity.

Standby pumps

It is emphasised that a pump breakdown can be expensive if the works become flooded and damage occurs. Therefore 100% reserve should always be immediately available.

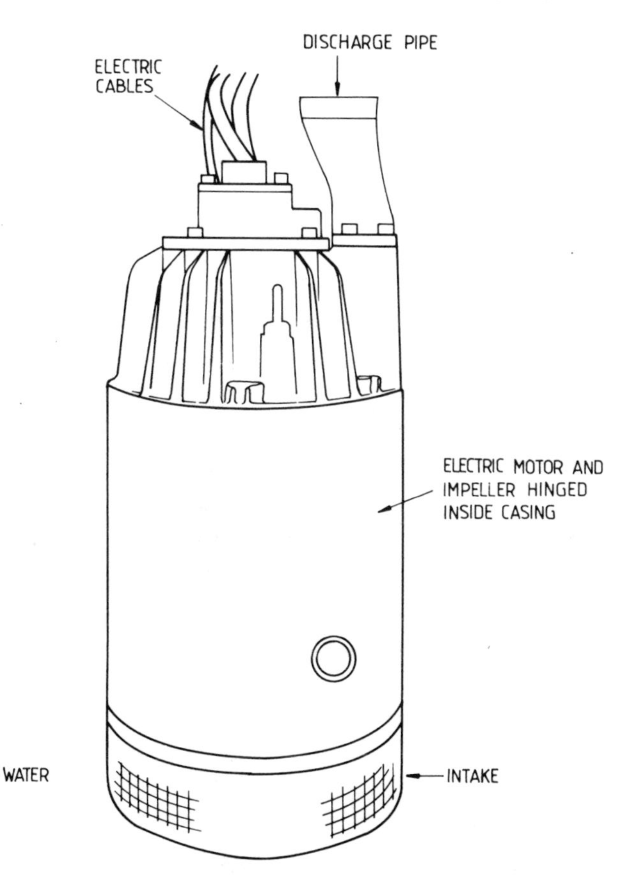

Fig. 6.36 Submersible centrifugal pump

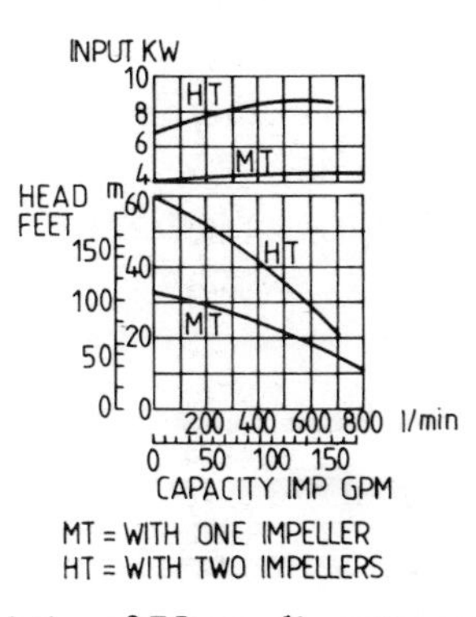

Fig. 6.37 Output characteristics of 75 mm diameter connection (3 kW) submersible pump

References

Bowen, R. (1980) *Ground Water.* Applied Science Publishers, London.

Casagrande, L. (1952) 'Electro-Osmotic stabilization of soils.' *Journal of the Boston Society of Civil Engineers*, U.S.A.

Cedergren, H.P. (1977) *Seepage Drainage and Flow Nets.* Wiley Interscience, NY, U.S.A.

Gibson, U.P. & Singer, R.D. (1971) *Water Well Manual.* Premier Press, Berkeley, Cal., U.S.A.

Huisman, L. (1972) *Groundwater Recovery.* Macmillan, London.

ITT Flygt Pumps Ltd. (1983) Trade literature, Colwick, Nottingham.

E.E. Johnson Inc. (1966) *Groundwater and Wells*, Saint Paul, Minnesota, U.S.A.

Jureka, W. (1975) *Dewatering.* Technical University of Aachen, W. Germany.

Kinori, B.Z. (1970) *Manual of Surface Drainage Engineering.* Vol. **1**, Elsevier, Holland.

Luthin, J.N. (1978) *Drainage Engineering.* 2nd Edn. Krieger.

Millar's Well Point International Ltd. (1982) Trade literature. Bishops Stortford, Herts.

Moller, B. (1957) 'Vakuumverfahren und die Grundwasserabsenkung nach der Well-point-Methode.' *Baumaschine and Bautechnik*, **4** (1), 19–27.

Powers, J.P. (1981) *Construction Dewatering.* Wiley & Son, N.Y., U.S.A.

Sichardt & Krieleis (1930) *Grundwasser Absenkungen bei Fundierungsarbeiten*, Berlin.

Small Wells Manual. 1969 Department of State, Agency for International Development, Washington D.C., U.S.A.

Sykes Ltd. (1982) Trade literature, Union House, Southwark Street, London.

Terzaghi, K. & Peck, R.B. (1967) *Soil Mechanics in Engineering Practice.* Wiley & Son, N.Y., U.S.A.

Thiem, G. (1906) *Hydrologische Methoden*, J.M. Gephardt, Verlag, Leipzig.

Wickham Engineering Ltd. (1982) Pumps and associated literature, Crane Mead, Ware, Herts.

CHAPTER SEVEN

Grouting

Grouting is the term used to describe the method of injecting a fluid substance into rock fissures or into a soil either to provide improved stability or to reduce permeability. The grout substance can be selected from a range of materials such as cement, clay suspensions in water, chemical solutions, or even emulsions of bitumen and water (Fig. 7.1).

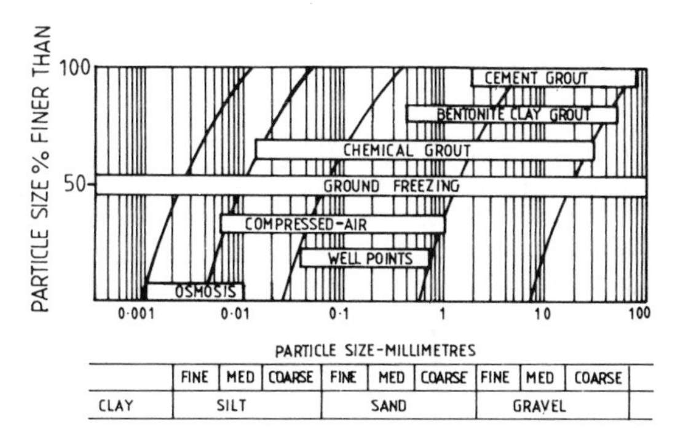

Fig. 7.1 Soil particle size and the choice of grouting or de-watering method

The grout is selected to meet the requirements of a particular situation and different techniques have been developed for the various grout types. Grouting is an expensive process and therefore every effort should be made to investigate the nature of the ground before the work commences.

Grouting applications

Grouting is used both for temporary and permanent works and has applications in:

(i) Sealing pockets and lenses of permeable or unstable soil or rock prior to excavation of a tunnel heading (Fig. 7.2) or alternatively grouting a stratum from ground level (Fig. 7.3).

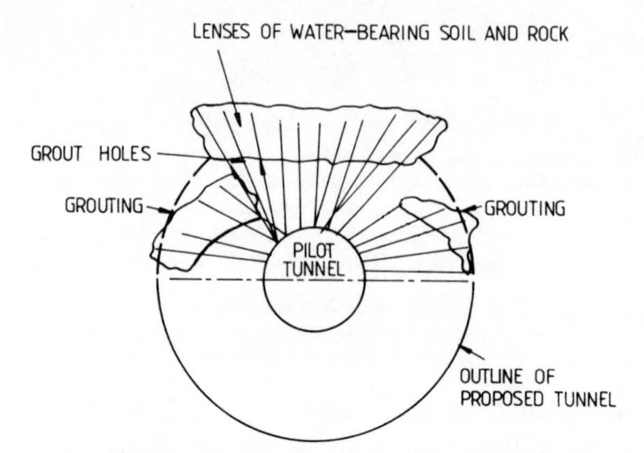

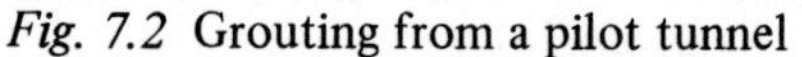
Fig. 7.2 Grouting from a pilot tunnel

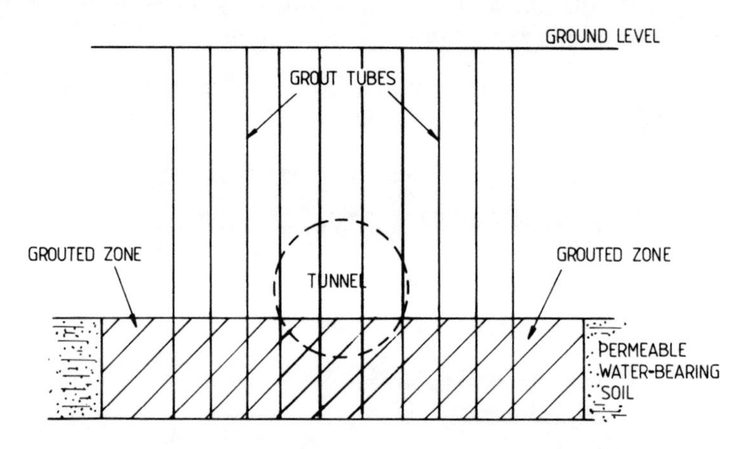

Fig. 7.3 Grouting a tunnel line from the surface level

(ii) Sealing the base of an excavation, coffer-dam or caisson founded on a permeable stratum (Fig. 7.4).

(iii) Grouting in ground anchors (Fig. 7.5) for sheet pile walls, concrete pile walls, retaining walls, stabilising rock cuttings, tunnels, etc.

(iv) Repairing: – (a) the ground underneath a foundation may be strengthened with the injection of a suitable grout

(b) a new damp proof course can be formed in a layer of brickwork by using a chemical grout

(c) cracks and structural defects on building masonry can be filled

(d) pavements and sunken slabs can be raised, etc.

(v) Filling the void between the lining and rock face in tunnel works (Fig. 7.6).

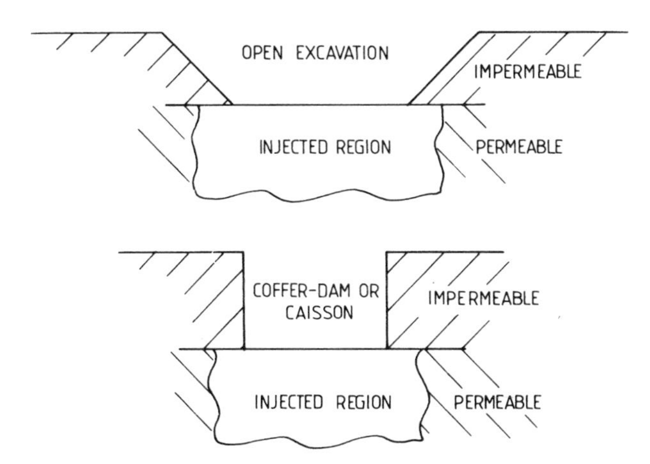

Fig. 7.4 Grouting a permeable stratum at the base of an excavation

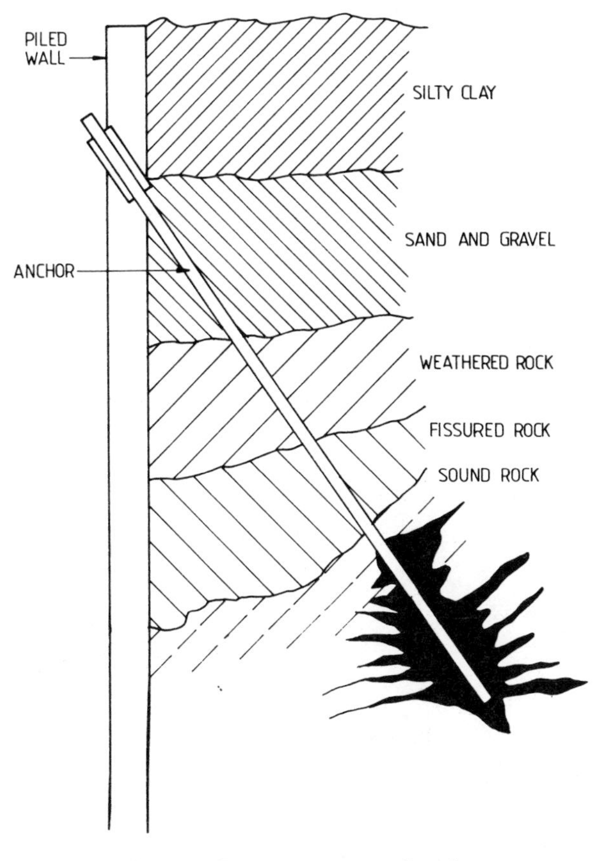

Fig. 7.5 Grouting in ground anchors

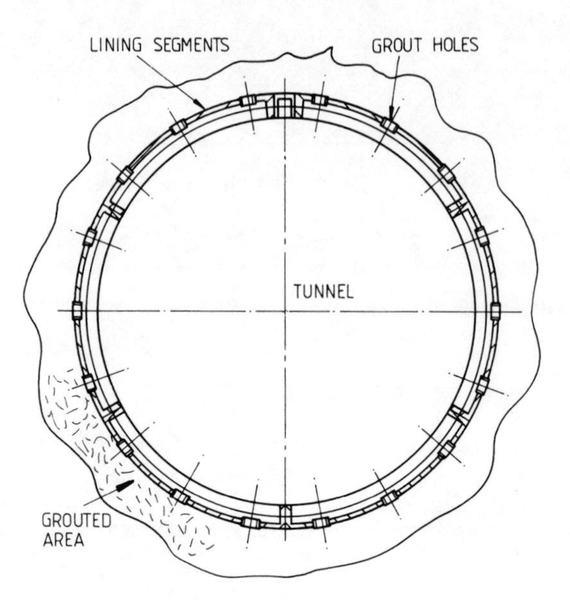

Fig. 7.6 Filling voids behind tunnel lining

(vi) Forming a 'grout curtain' in layers of permeable strata below a dam and so effectively sealing off any flow from the storage side. This method may be applied either before construction takes place or alternatively to effect repairs to an existing structure (Fig. 7.7).

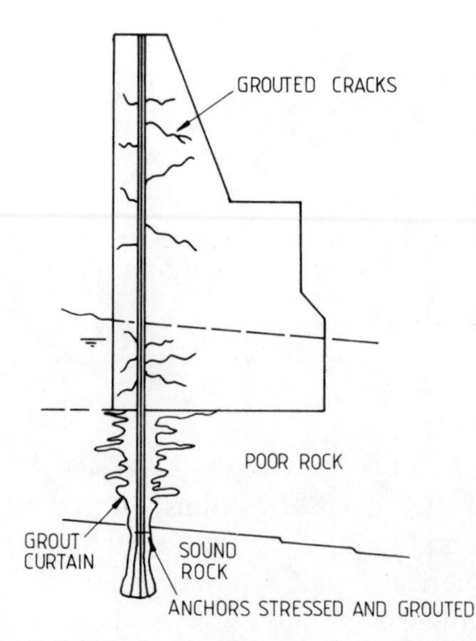

Fig. 7.7 Forming a grout curtain below a dam

(vii) Grouting up the tendons in prestressed post tensioned concrete.
(viii) Sealing the gap with a non-shrinkable grout between the surface of a concrete foundation and the base plate of a stanchion, etc.
(ix) Producing mass concrete structures and piles by introducing cement after the aggregate is in position (Fig. 7.8).

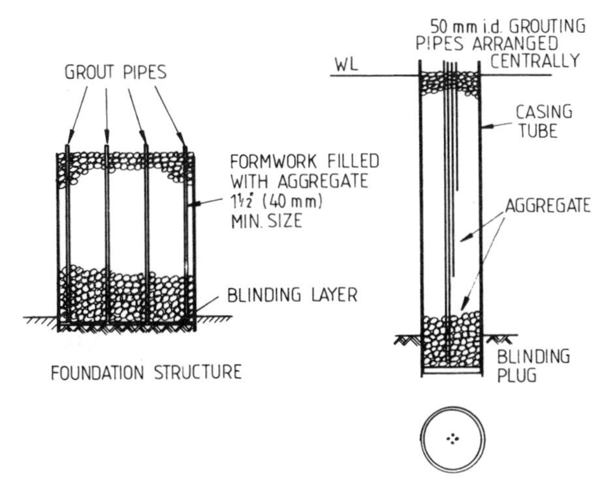

Fig. 7.8 Producing foundations and piles (Colcrete)

Principles of grouting

The principle of grouting is to introduce a substance to fill the voids in a soil or the fissures in a rock by pumping fluid down a small diameter tube placed in a drillhole. The depth of the boring controls the thickness of the layer to be grouted and as each stage is completed the borehole is lengthened and a further layer is grouted, and so on until the design depth is reached. (However, modern developments allow the borehole to be made to the full depth in one operation, with subsequent use of packers to obtain stage grouting; this technique and others are discussed later.)

Penetration of grout

If the medium into which grout is to be injected can be assumed to have equal permeability in all directions, then for a Newtonian fluid (see Fig. 7.16) the radius of penetration of the grout into the soil is assumed to be spherical spreading out from the point of injection (Fig. 7.9). The mathematical analysis according to Maag is as follows:

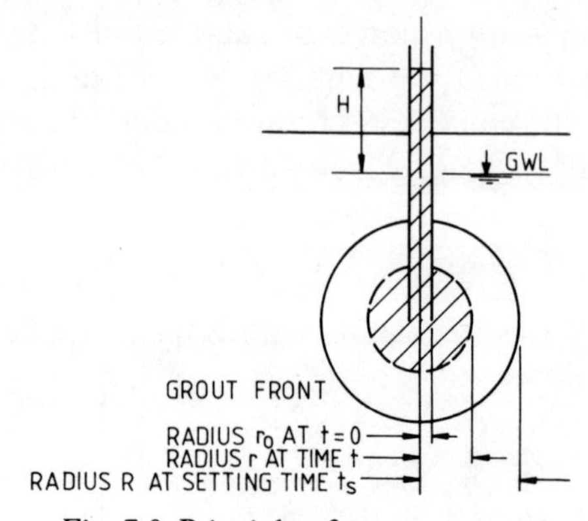

Fig. 7.9 Principle of grout penetration

$$q = A \times v$$

where A is the surface area of a sphere of grout of radius r, and
v is the velocity of flow of the grout.

From Darcy's Law

$$v = -k_g \frac{dh}{dr}$$

where k_g is the coefficient of permeability of the soil with a grout flow in length/unit time. $-\frac{dh}{dr}$ is the rate of change of fluid head.

Thus from Fig. 7.9,

$$q = -4\pi r^2 k_g \frac{dh}{dr}$$

$$\text{or} - \int_H^o dh = \int_{r_o}^R \frac{q}{4\pi r^2 k_g} \times dr$$

By integration

$$-h = \frac{q}{4\pi r k_g} + C$$

when $r = r_o$ the radius of the grout tube then $h = H$ the effective head of grout

therefore $C = -\frac{q}{4\pi r_o k_g} - H$

when $r = R$, the limiting radius of grout penetration $h = 0$.

therefore $C = \frac{-q}{4\pi R k_g}$

Substituting for C, then

$$\frac{q}{4\pi R k_g} = H + \frac{q}{4\pi r_o k_g}$$

$$q = \frac{4\pi k_g H}{\left(\frac{1}{R} - \frac{1}{r_o}\right)} \tag{7.1}$$

But if the fluid increases the radius of the sphere by dr in time dt, and n is the porosity of the soil (i.e. ratio of voids to total volume), then

$$q = 4\pi r^2 \frac{dr}{dt} \times n$$

Integrating,

$$\int_{r_o}^{R} nr^2\, dr = \frac{q}{4\pi} \int_{o}^{T} dt$$

$$\frac{nr^3}{3} = \frac{qt}{4\pi} + C$$

when $r = 0, t = 0$ and so $C = \dfrac{nr_o{}^3}{3}$

when $r = R, t = T$ and so $C = \dfrac{nR^3}{3} - \dfrac{qT}{4}$

Substituting for C then

$$\frac{qT}{4\pi} = \frac{n}{3}(R^3 - r_o{}^3) \tag{7.2}$$

Substituting for q and solving (7.1) and (7.2)

$$\frac{4\pi k_g H}{\left(\frac{1}{R} + \frac{1}{r_o}\right)} = \frac{n}{3}(R^3 - r_o{}^3)$$

but $\dfrac{1}{R}$ is negligible compared to $\dfrac{1}{r_o}$

and

$$k_g = k_w \frac{\eta_w}{\eta_g} \times \frac{\gamma_g}{\gamma_w} \tag{7.3}$$

where k_w = coefficient of permeability of soil with water flow
η_w, η_s are the respective dynamic viscosities of water and grout, e.g. in Newton-seconds per m^2
γ_w, γ_g are the respective unit weights of water and grout e.g. in N/m^3
Therefore

$$R = \left(\frac{3 \times r_o}{n} \times k_w \times \frac{\eta_w}{\eta_g} \times \frac{\gamma_g}{\gamma_w} \times H \times T + r_o{}^3\right)^{\frac{1}{3}} \qquad (7.4)$$

or

$$R = \left(\frac{3 \times r_o}{n} \times k_w \times \frac{\nu_w}{\nu_g} \times H \times T + r_o{}^3\right)^{\frac{1}{3}} \qquad (7.5)$$

where ν represents kinematic viscosity.

In practice the viscosity of the grout changes with time and eventually gels. Thus it can be seen from equation (7.4) that if all the other factors remain constant, the radius of penetration R is reduced as the viscosity of the grout increases.

However, it is also apparent that the pumping pressure (or hydraulic head H) may be increased to counteract this effect. Other influential factors are the grout density, soil porosity, soil permeability, and the diameter of the grout pipe. The soil permeability in particular is affected by the particle size, arrangement and distribution, the continuity of pores and formation stratification. Furthermore, the permeability in the horizontal direction generally exceeds that in the vertical and the resulting injected volume tends to be cylindrical (Fig. 7.10) rather than the theoretically assumed sphere.

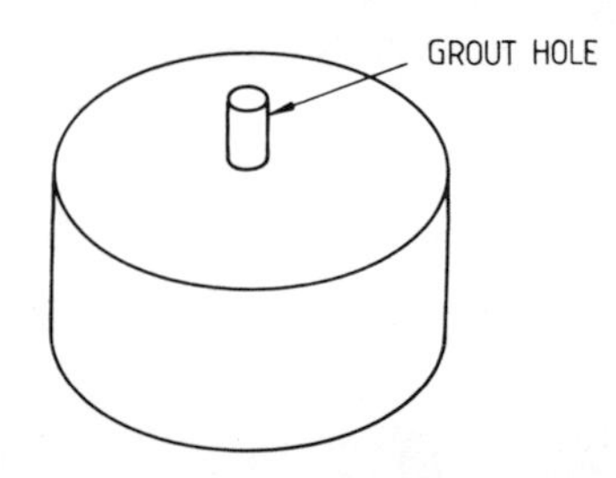

Fig. 7.10 Effects of soil permeability on grout penetration

The presence of ground water, non uniform soils, fissures in rock, etc. often cause distortions to the theory and therefore the injection pattern, pumping pressures, etc. must be tempered by experience based on information obtained from a full site investigation.

Pre-grouting site investigation

To obtain satisfactory grouting of the soil or rock strata it is essential to carry out a full site investigation before commencing operations. This may be expensive, but without it the extent of required grouting work could only be guessed. The site investigation may be undertaken separately from the grouting contract and in many cases this is preferable in order that full and proper testing can take place, unhindered by influences such as the contractors' need to perform the grouting task in the most economic time programme. The investigation should include a geological survey, and investigation drilling.

Geological survey

This consists of examining the general geology of the area, using mapping methods, supplemented by exploratory drilling to establish fissure systems, faults, folds, etc. The final geological map should provide information on the extent of soil and rock formations, zones of weakness, the dip and strike, etc. of the strata.

Site investigation

The site investigation involves a second and more detailed exploration of the precise location where grouting is to be concentrated, and is used to determine the criteria for estimating the type, quantity and extent of the grouting necessary. Core samples are taken during borehole drilling for analysis in the laboratory to determine rock or soil strength, grain size, permeability, porosity, etc. An important aspect of the investigation is in situ permeability tests on the strata, which yield more reliable data than tests made on individual samples in the laboratory, and therefore facilitate more precise estimation of the required pumping pressures, grouting pattern and type of grout.

In situ permeability tests

The approximate permeability of a stratum may be determined by several methods, using either a constant or variable head of water. The method involves sinking a borehole through the stratum. The drill stem of the boring equipment is then used to maintain a head of water above the ambient pore pressure in the ground.

The rate of flow at different time intervals is measured, to obtain a projection of the steady state rate of flow. In the case of a constant head, the coefficient of permeability (k) is subsequently calculated from a formula of the form

$$k = \frac{q}{H} \times \text{factor } (F)$$

where q is the steady state rate of flow, and
H is the hydraulic head

The factor (F) is dependent upon the shape of the exit and values can be obtained either by reference to Hvorslev or Gibson.

It is usual to log the soil at 1 m intervals over the entire length of borehole as shown in Fig. 7.11. This is achieved with the aid of two packers (i.e. plugs) as illustrated in Fig. 7.12.

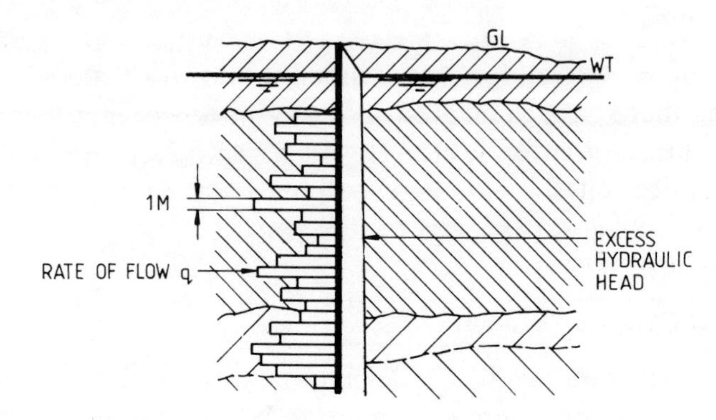

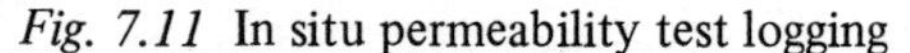

Fig. 7.11 In situ permeability test logging

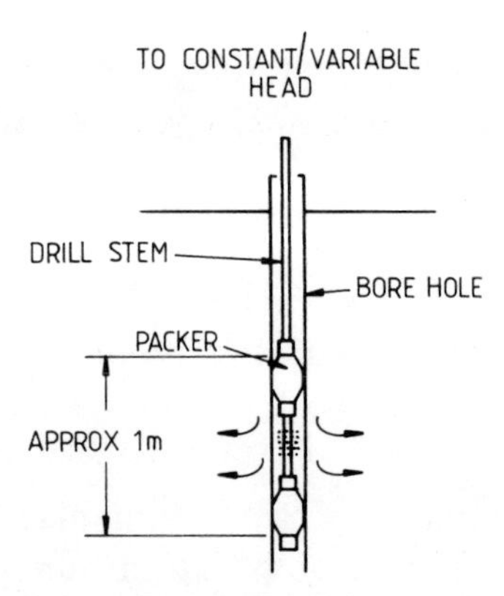

PACKERS FOR PERMEABILITY MEASUREMENT

Fig. 7.12 In situ permeability testing method

In rock strata, because of the presence of fissures, it is common practice to determine an equivalent permeability by the Lugeon test. Units of Lugeon are the flow in litres per minute absorbed by 1 m of drill hole at an injection pressure of 1 N/mm^2 above ambient pore pressure.

One unit of Lugeon is equivalent to a conventional permeability (k_w) of approximately 1×10^{-4} mm/s.

With in situ values of permeability (k_w) determined for water, the equivalent value for a grout (k_g) may be calculated from eqn. (7.3).

By assimilating other data from the site investigation, the theoretical extent of grout penetration may be ascertained from eqn. (7.4).

Specification for grouting

The results of a borehole permeability log indicate the zones to be grouted, the type of grout to choose, and the likely extent of grout penetration under a given pressure. Practical experience, however, has demonstrated that spacings of grout holes in different soil types should be similar to those given in Table 7.1, but this will very much depend upon the viscosity of the grout being used.

Table 7.1 Grout hole spacing

Coefficient of permeability (soil-water) (mm/s)	Grid spacing (m)	Soil types
>1	6 m	Fissured rocks
1 to 1×10^{-1}	3 m	Medium/coarse sands and gravels
$<10^{-1}$	0.5 to 1 m	Fine sands

Note: Generally after adequate grouting, permeability can be reduced below 1×10^{-3} mm/s.

Once the length, breadth and depth of the zone to be grouted and the grout hole spacing are fixed, the quantity of grout required can be calculated. The total volume required is dependent upon the voids ratio of the mass to be injected. However, exact estimates are difficult to obtain because of the necessity of altering viscosity, pumping rates and pressures to accommodate the unknown influencing factors such as ground water conditions, fissures, etc. during the execution of the work. However, for a simple example of a grouted sphere of 1 m radius of voids ratio 0.6, the volume of grout required

$$= \frac{4}{3} \times \pi \times 1^3 \times \frac{0.6}{1.6} = 1.57\,\text{m}^3 \text{ or } 1570 \text{ litres.}$$

It can be seen in Fig. 7.13 that flow from the primary holes travels out on a radial front while flow from secondary and tertiary holes fills the spaces between adjacent primary injections. Estimates of grout volumes should be made on this basis.

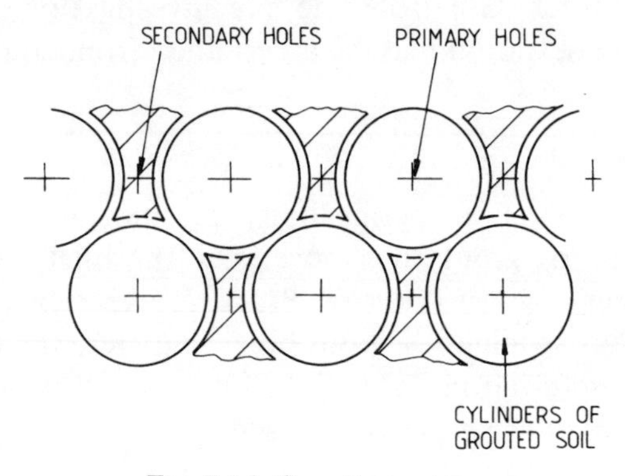

Fig. 7.13 Grouting pattern

Grout design

Grout is injected into soil, rock or other material to:

(i) Reduce the flow of ground water
(ii) Increase the strength of the material

To obtain these effects the grout must set and harden, but not so quickly that pumping becomes impracticable. Additives are therefore frequently required to control the viscosity (gelling time) and subsequent rate of hardening. The viscosity and strength of grouts generally increase with respect to time as shown in Fig. 7.14. For example, cement- or clay-based grouts may take 28 days or more to reach almost full strength. Most important, however, during injection is the fluid viscosity, which gradually changes as chemical reactions take place to transform the grout from a liquid into a gel (i.e. initial set). Some grouts react in minutes, while for others maximum viscosity is only obtained after about 24 hours.

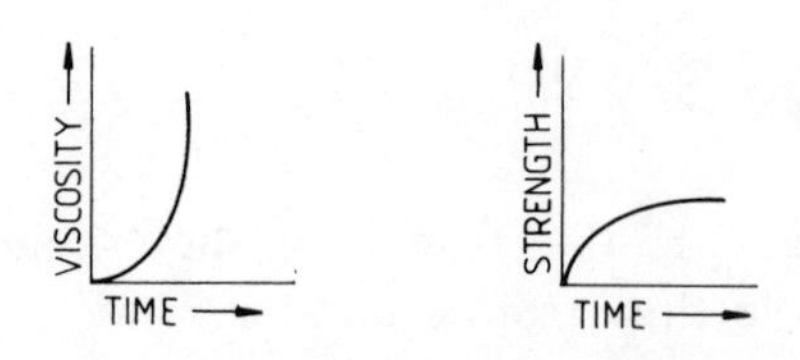

Fig. 7.14 Variation in grout strength and viscosity with time

Viscosity

When a fluid is disturbed its various parts move at different velocities, which come to rest when the disturbance is removed. For example, for

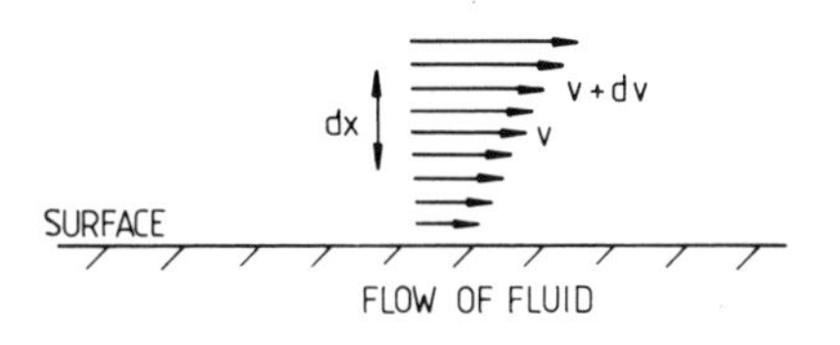

Fig. 7.15 Principles of viscosity

a liquid in motion on the plane solid surface shown in Fig. 7.15, the layer in contact with the surface is at rest, while the layers above move at increasing velocities to produce a velocity gradient $\frac{dv}{dx}$. According to Newton there is a shearing stress between any two layers, and the relationship between this stress (f) and the velocity gradient is called the coefficient of dynamic viscosity i.e.

$$\frac{f}{\frac{dv}{dx}} \text{ (Expressed as units of poise or } \frac{1}{100} \text{ poise, i.e. cP)}$$

$$1\,\text{cP} = 10\,\text{N} - \text{s/m}^2.$$

The viscosity of water at 15°C is approximately 1 cP.

If change in viscosity caused by hardening or temperature is ignored, then a plot of rate of shear $\frac{dv}{dx}$ against shear stress produces a straight line, as shown in Fig. 7.16, for a purely viscous liquid such as water or a thin chemical solution. Such liquids are known as Newtonian liquids.

For suspensions in water such as cement or clay, and some chemical precipitates, the curve no longer passes through the origin and a yield value is exhibited. Such suspensions are called Bingham fluids and

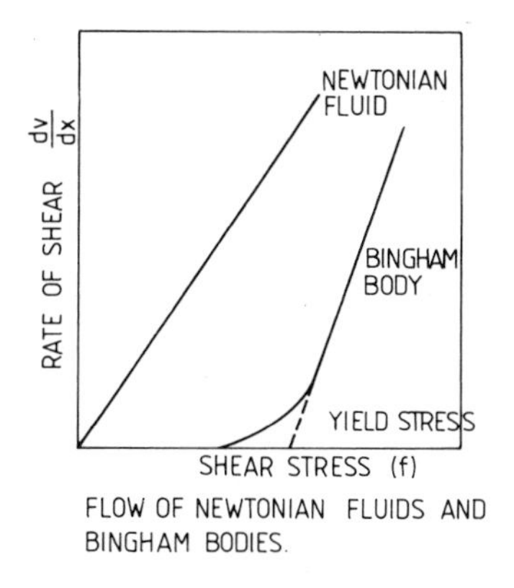

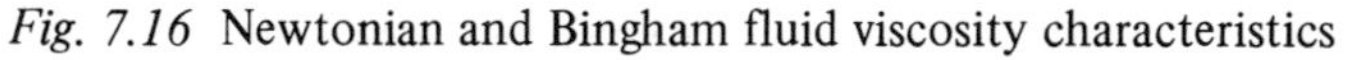

Fig. 7.16 Newtonian and Bingham fluid viscosity characteristics

extra force must be applied to overcome the inherent shear strength. Some Bingham fluids, e.g. bentonite suspensions, are thixotropic, and behave like a fluid when in movement, but form a 'jelly' when stationary. Usually the transformation is quite reversible and may be repeated.

Grouts which exhibit the above characteristics fall into the following categories:

(a) suspensions of solid particles in water, such as clays, cements, bentonite, plaster, PFA, lime, etc.
(b) emulsions, such as bitumen in water
(c) solutions, which react after injection to form insoluble precipitates

The principles to follow in choosing the grout are:

(i) the grout must be able to penetrate the voids of the mass to be injected (e.g. a cement grout or flocculating chemical grout may get filtered in a fine sand).
(ii) the grout should be resistant to chemical attack when in place.
(iii) the grout should be able to develop sufficient shear strength to withstand the hydraulic gradient imposed during injection and on flowing ground water. Most grouts are suitable to meet this latter consideration.

A wide selection of grouts is available and Table 7.2 sets out recommendations for applications in various situations (see Fig. 7.1).

Table 7.2 Grout types and applications

PFA	Suspensions	Mass filling in very coarse soils and rock fissures
Cement	Suspensions	Mass filling in very coarse soils and rock fissures plus ground strengthening
Clays	Suspensions	Mass filling in medium coarse soils and impermeability improvement
Clay/cement	Suspensions	Similar to clays, plus added strength
Emulsions		Impermeability improvement
Solutions, single shot		Permeability and/or strength improvement in medium coarse soils
Solutions, double shot		As for single shot, with additional control of gel time. Also suitable in fine soils

Table 7.3 Physical properties of PFA grouts

Type of grout	Water/ Solids	PFA/ cement	28 day crushing strength (N/mm^2)	Density (kg/m^3)	Setting time (h)	Viscosity (cP)
PFA	0.5	–	0.2	1100 - 1500	24	400
PFA/Cement	0.5	20	0.8	1100 - 1500	24	500
PFA/Cement	0.5	10	2	1100 - 1500	24	750
PFA/Cement	0.5	5	4	1100 - 1500	24	1000

Grout types

PFA grout (*Table 7.3*)

Pulverised fuel ash reacts with lime and water to produce a stable cementitious material. The particle size is similar to cement and therefore is mainly used in highly permeable materials (i.e. greater than 1 mm/s) such as fissured rock, gravels and coarse sands.

A water:solids ratio of about 0.50 produces a free-flowing grout, but at 0.35 a thick slurry is formed which is a little too viscous for optimum pumping.

When set, the hardened grout has good resistance to sulphate attack and shrinkage is negligible.

Crushing strength can be improved by adding cement, and PFA-cement ratios by dry weight of 5:1 to 20:1 are common.

Cement grout

The average specific surface area (i.e. fineness) of ordinary portland cement (OPC) particles is approximately 30 mm^2 per gram, the minimum being 20 mm^2/g. Therefore like PFA, OPC grouts are suitable only in fissured rocks, gravels and coarse sands.

The water:cement ratio may be varied from about 0.6:1 to 3:1 depending upon the ground conditions and required strength.

Rapid hardening cement is finer than OPC and produces a quicker setting time and high early strength, and therefore may be preferred to OPC in ground with high flowing water.

High alumina cement also has rapid strength gain and offers good resistance to attack by sulphates and dilute acids. Supersulphated cement has a fineness of about 60 mm^2/g and is therefore suitable for penetrating finely fissured rocks.

In very coarse materials and fragmented rocks, sand is often added as a filler as dictated by the pumping conditions.

Cement grouts are mainly selected for ground strengthening and after 28 days, crushing strengths of up to 60 N/mm^2 can be obtained depending upon the water:cement ratio.

Clay grouts (*Table 7.4*)

Table 7.4 Physical properties of clay grouts

Type of grout	Water/ solids	Clay/ cement	Density (kg/m^3)	Crushing strength (N/mm^2)	Setting time (h)	Comments
Clay-chemical	20	–	1100	–	Varies	Fluid
Clay-cement	7	0.5	1100	0.2	24	Thin slurry
Cement-bentonite clay	0.3	0.05	1600	5	24	Thick slurry

Clay is a complex compound made up of minute mineral particles less than 0.002 mm in diameter, and is thus suitable for injection into medium coarse sands and other soils with a permeability of 1 to 10^{-1} mm/s. Clay grout behaves like a Bingham fluid and gels when undisturbed. However, poor strength characteristics are exhibited and so it is mainly used to reduce permeability.

Kaolinite or Illite based clays produce low viscosities and are therefore preferred as filler grouts. Viscosity can be reduced by the addition of a dispersing chemical such as sodium phosphate. A rigidifying agent such as sodium silicate can then also be included to improve the shear strength when set. However, the final shear strength is likely to be poor and the material can be displaced by a hydraulic gradient of 3 – 4 units.

Clay grouts tend to flocculate when in contact with acid water and may get filtered if the soil is too fine.

Cement may be added to clay to increase its shear strength but because of the relatively large particle size of cement it is limited to use in coarse grained soils or fissured rocks. This type of mixture is called a clay–cement grout, because of the predominance of the clay. Where the clay is a smaller proportion of the total, the grout is referred to as a cement–clay grout. In this latter case bentonite clay is sometimes added in small quantities to reduce the tendency of the cement particles to settle out before full penetration is reached.

Bitumen

Hot bitumen at about 150°C is an effective grout in highly permeable soils containing flowing water which might wash away clay or cement grouts during injection. As hot bitumen makes contact with the colder ground water, its outside surface skins over, while the inside stays hot and fluid and remains capable of being injected over extensive distances, especially in fractured or fissured rock strata. When the bitumen solidifies, the void is completely filled to provide an effective water seal.

Alternatively, bitumen emulsions containing 50 – 60% bitumen, 1% emulsifier, and water are suitable for application in fine to medium sands. A coagulant is often included to precipitate the bitumen, the resulting particles gradually building up in the soil pores under the action of pumping. Bituminous grouts are less common today, as developments in chemical grouts prove more economical.

Chemicals

Like cements, chemical grouts rely on a chemical reaction to produce a material of continuous structure when set. But, unlike cement particles suspended in water, a chemical grout is formed from two separate solutions which react together to produce the gel, which subsequently hardens. The effect can either be obtained by injecting the two fluids one after the other, known as 'two shot' grouting, or alternatively by a single injection of a chemical containing an accelerator to induce gelling.

Two shot grouting

This method was developed by Joosten and consists of an initial injection of sodium silicate followed by one of calcium chloride. The gelled compound behaves like a Bingham fluid, and because of its relatively high viscosity (100 cP) compared to thin suspension grouts, the grout hole spacings must be close together, 700 mm being typical. As a consequence, high pumping pressures are necessary. The method is limited to use in medium-coarse sands (permeability 10 – 1 mm/s or higher). The hardened gel has a strength comparable with cement grout, i.e. approximately 8 N/mm^2 after 28 days.

Single shot (*Table 7.5*)

To overcome the viscosity disadvantages of the two shot method, Guttman modified the Jooster process by diluting the sodium silicate solution with sodium bicarbonate solution (resulting viscosity 20 cP and lower strength), but more recent developments have concentrated

Table 7.5 Physical properties of examples of single shot chemical grouts

Type	Viscosity (cP)	Gel time (min)	Specific gravity	Application (soil permeability) (mm/s)	Crushing strength (N/mm^2)
Sodium silicate-sodium bicarbonate	1.5	0.1 - 300	1.02	10^{-1}	0.5
Chrome-lignin	3.0	5 - 100	1.10	10^{-1}	0.5
Sodium silicate-amide	5 - 50	5 - 300	1.10	10^{-1}	5
Resinous polymers	1.0 - 1.3	1 - 300	1.01	10^{-2}	0.5

on a 'one shot' method whereby the chemical additives slowly react with the sodium silicate to form a gel. Many varieties are now available, possessing a range of properties from low to high strength, viscosity, etc., which allows grouting of fine sands with a permeabilty of about 10^{-1} mm/s.

Other types of chemical grouts have been developed in recent years, for example, resin-based compounds can be used in very fine soils of permeability 10^{-2} – 10^{-3} mm/s.

Common examples of one shot chemical grouts are given in Table 7.5. However, it is emphasised that chemical grouting is more expensive than cement and clay grouting.

Grouting procedures

Hole pattern

Ideally, the spacing of the grout holes should be set out on a grid pattern such that the radius of penetration is sufficient to cause slight overlapping between adjacent holes. A second and subsequent half-size grid is then injected to fill the spaces between adjacent columns (see Fig. 7.13). A third, quarter-size grid is sometimes required to achieve an acceptable reduction in permeability.

Grout consistency

The grouting speed and rate of pumping are governed by the relationship

> Available power from pump $\propto$ pumping pressure $\times$ quantity of grout delivered per unit time.

Before grouting starts it is good practice to flush clean the drill hole with water and indeed to pump water into the soil or rock to clean out cavities or to lubricate the soil particles to aid grout flow.

At the commencement of pumping operations, grout is delivered at the maximum achievable rate. However, with suspension grouts pumping is normally started with a thin mix and the concentration gradually increased until the pressure begins to rise, with a corresponding fall in delivery rate. The viscosity is subsequently maintained at this consistency and the pressure increased up to the limit imposed by the overburden, to avoid heave and leakage to the surface (this should be calculated by reference to soil unit weight). The process is continued until the calculated volume of grout has been injected.

Methods of working

The stratum is divided into zones depending upon the permeability tests obtained during the site investigation and each zone is separately grouted. This may be achieved by several methods as follows.

Grouting from the bottom upwards (*Fig. 7.17*)

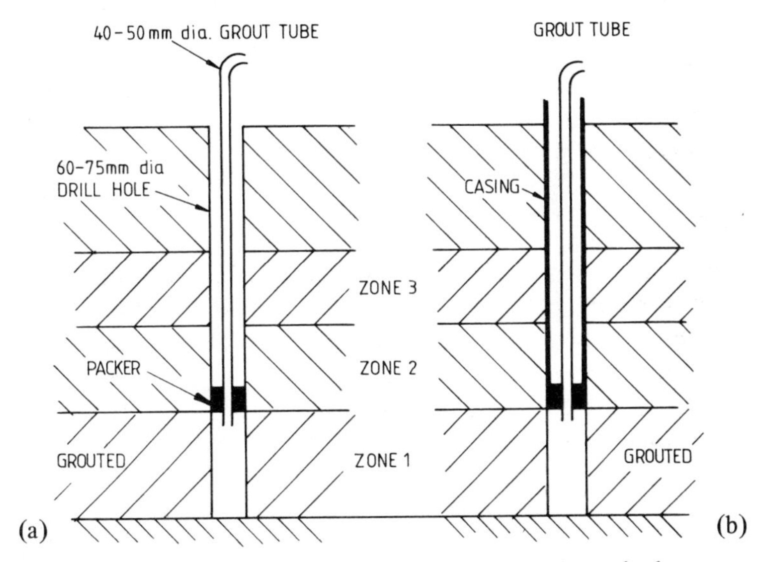

Fig. 7.17 Grouting from the bottom upwards method

A grout hole 50 – 75 mm diameter is drilled to full depth using one of the methods described in Chapters 3 and 5. In rock strata and rigid soils a self expanding packer (Fig. 7.18) is placed directly above the lowest zone and grout is pumped in. The packer is then raised to the next zone and the procedure repeated, thereby grouting the drill hole successively upwards. In soft or unstable soils the drill hole must often be lined with a casing to provide support and to provide a good seal between the packer and borehole walls. The casing is progressively raised with the packer as shown in Fig. 7.17(b). In fissured rock, a particularly permeable stratum can be isolated using a double packer (Fig. 7.19).

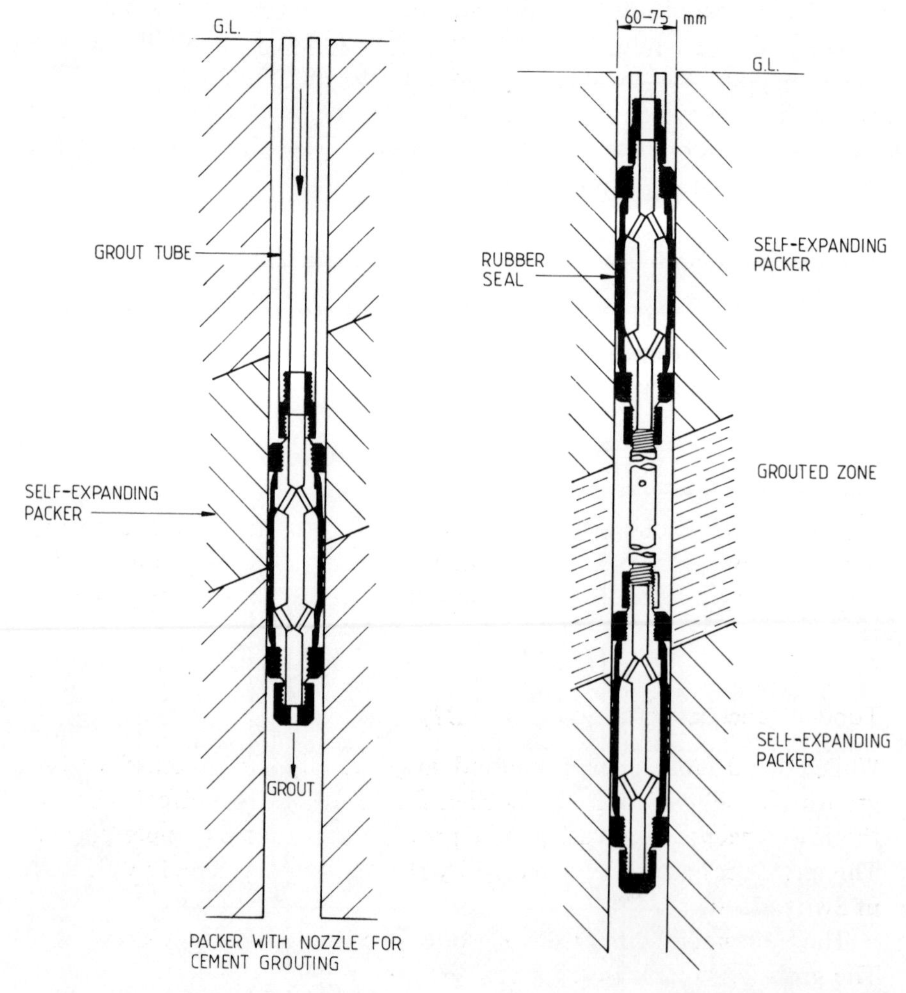

Fig. 7.18 Self-expanding packer

Fig. 7.19 Grouting an isolated stratum using self-expanding packers

Grouting from the top downwards

Grouting commences in the top zone, the drill hole is then deepened and the next zone grouted. The procedure is repeated until the full depth has been treated. This method is appropriate for shallow working, as the grouted zones then provide a seal and prevent leakages to the surface.

Circulation grouting (*Fig. 7.20*)

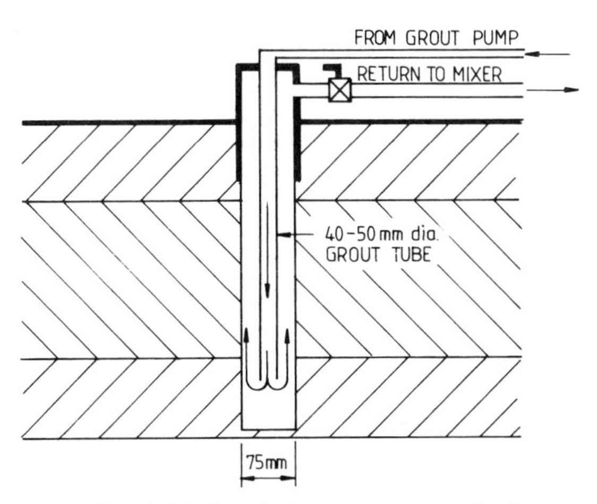

Fig. 7.20 Circulation grouting method

This method operates on the principle of grouting from the top downwards. A drill hole is bored to the depth of the first zone and grout is pumped down the grout pipe and returned up the drill hole. In this way, clogging is reduced. The hole is then deepened and the procedure repeated.

Tube-à-Manchette grouting (*Fig. 7.21*)

While the double packer method is practicable in rocks, for zonal grouting in alluvial soils it has been found difficult to effect a seal on the lower packer (the casing tube provides a seal for the upper packer). This problem has been overcome by the method developed by E. Ischy in Switzerland.

The equipment comprises a grout pipe located inside a sleeve pipe. The grouting zone is isolated between two packers and grout is pumped through holes in the sleeve pipe, set at about 0.3 m intervals, covered

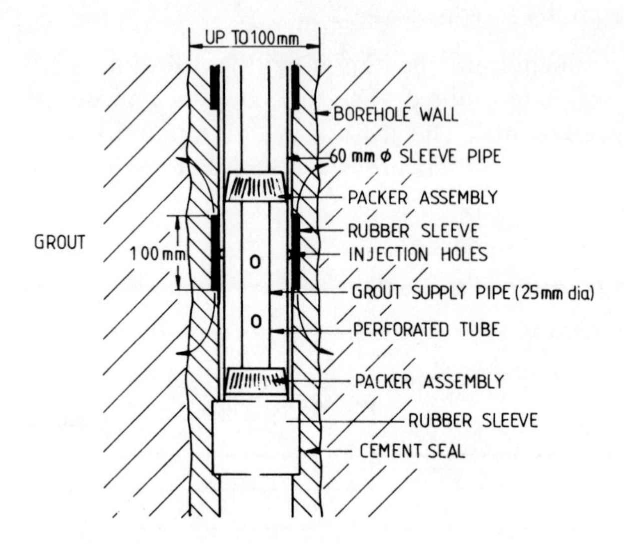

Fig. 7.21 Tube-à-Manchette grouting method

with a rubber band. The void between the borehole wall and sleeve pipe is filled with a thin cement layer to form a seal. During grouting, the grout is forced locally through the seal and enters the soil at the required depth.

Point grouting

In shallow work, perhaps 10 – 12 m deep, grouting is injected at predetermined levels from a driven lance (Fig. 7.22). Further grouting may

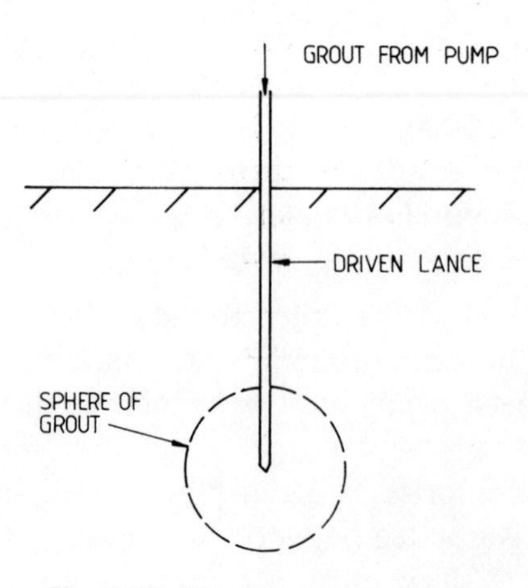

Fig. 7.22 Point grouting method

take place as the tube is withdrawn. The technique is commonly used with chemical grouts, where close spacings are required.

Jet grouting (*Fig. 7.23*)

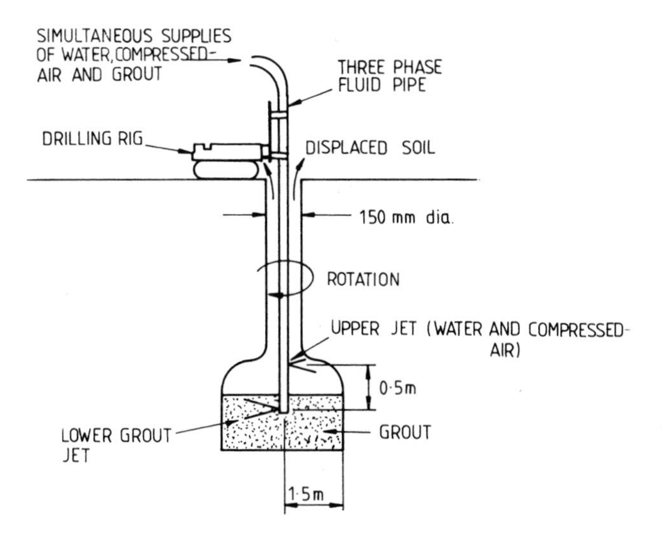

Fig. 7.23 Jet grouting method

Jet grouting was originally developed in Japan, and has recently been introduced into Europe. Unlike conventional injection methods, jet grouting is used as a replacement technique. Soils ranging from silt to clay and weak rocks can be treated. The method requires a borehole approximately 150 mm diameter, into which a drill pipe is lowered. The pipe is specially designed to convey simultaneously pumped water, compressed air and grout fluid. Near the bottom end of the pipe, two nozzles are located 500 mm apart. The upper nozzle (1.8 mm diameter) delivers water at about 400 bar, surrounded by a collar of compressed air at 7 bar to produce a cutting jet. The lower nozzle (7 mm diameter) delivers the grout at approximately 40 bar. The grouting action requires the stem to be slowly raised, whereby the excavated material produced from the jetting action is replaced by the grout and forced to the surface. The jet of water has an effective reach of about 1.5 m and thus by rotating the stem, a column of replaced earth may be formed.

Jetting data

Columns	0.5 - 2 m diameter
Depth	In excess of 15 m

Grouting rates	180 - 200 *l*/min
Water	70 *l*/min at 400 bar
Withdrawal rate	50 mm/min

Table 7.6 Jet grouting with a water cement slurry

Soil	W/C ratio	Compressive strength (N/mm^2)	Final co-efficient of permeability (mm/s)
Granular	1	5 - 10	10^{-5} - 10^{-8}
Cohesive	1	1 - 5	10^{-5} - 10^{-8}

Grouting equipment (*Fig. 7.24*)

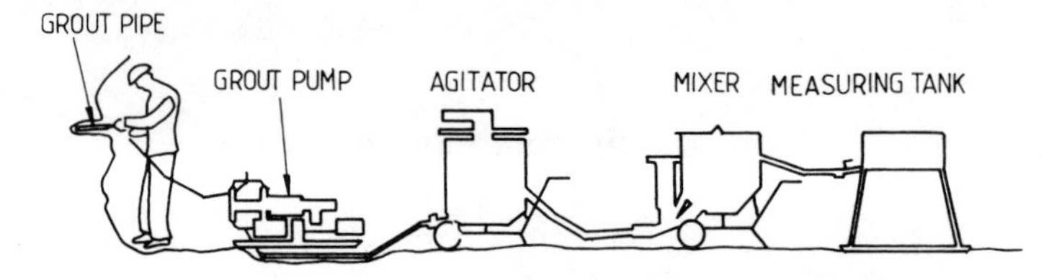

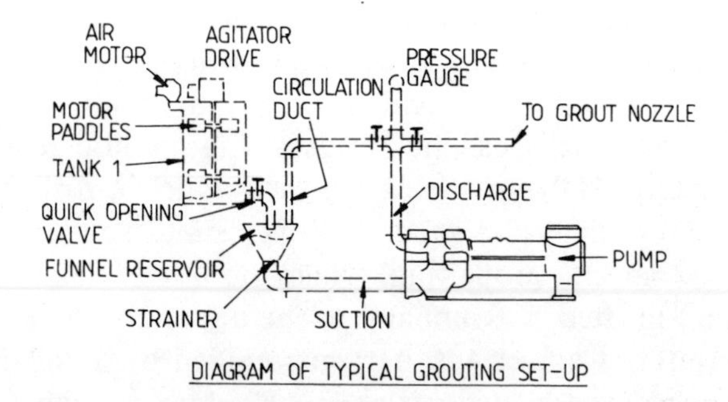

Fig. 7.24 Grouting equipment

The basic items required are:

(i) Measuring tank – to control the volume of grout injected
(ii) Mixer – to mix the grout ingredients
(iii) Agitator – to keep the solid particles in suspension until pumped (not required for chemical grouts)
(iv) Pump (usually of the piston or diaphragm type) – e.g.

	Pressure (N/mm^2)	*Delivery* (l/min)
Small pump	0.8	120
	1.5	45
Large pump	3.5	450
	10	130

(v) Grout pipe, packers, casing tubes, flow meters, pressure gauges, etc.

Final permeability tests

After completion of grouting, further boreholes may be drilled and core samples taken and examined for grout penetration. Permeability of the treated ground may be determined from water tests conducted in a similar manner to the tests carried out during the site investigation.

Bibliography

Atlas Copco Company (1983) Cement grouting, Stockholm, Sweden.

Bowen, R. (1975) *Grouting in Engineering Practice*, Applied Science Publishers Ltd., London.

Brockett, R.W. (1974) 'Survey of modern grouts and grouting'. *Structural Engineer*, 52, May.

Clough, G.W. (1975) 'A report on the practice of chemical stabilisation around soft ground tunnels in England and Europe'. Dept. of Civil Engineering, Stamford University, USA.

Colcrete Ltd. (1983) The Colcrete process. Rochester, Kent.

Gibson, R.E. (1966) 'A note on the constant head test to measure soil permeability in-situ'. *Geotechnique*, Vol. 16, No. 3, pp.256–259.

Glossop, R. (1961) 'The Invention and Development of Injection Processes', Part I 1802–1850 *Geotechnique* **X** 3, Sept. 1961, Part II 1850–1960 *Geotechnique* **XI** 4, Dec. 1961.

Glossop, R. & Sichy, E. (1962) 'An introduction to alluvial grouting'. *Proceedings of the Institution of Civil Engineers*, **21**, pp.449–474.

'Grouting design and practice' (1969) *The Consulting Engineer*, (October). Construction Publications Ltd.

Health and safety aspects of ground treatment materials (1982). CIRIA Report No. 95.

Hvorslev, M.J. (1951) 'Time lag and soil permeability in groundwater observations', Bulletin No. 36, Waterways Experimental Station, Corps of Engineers, U.S. Army.

Karol, R.H. (1960) *Soils and Soil Engineering*, Prentice-Hall Inc., London.

Maag, E. (1938) 'Über die Verfestigung und Dichtung des Baugrundes (Injektionen)'. Course notes in mechanics. Zurich Technische Hochschule.

Perrott, W.E. (1965) 'British Practice of Grouting applied to Granular Materials'. ASCE Water Resources Engineering Conference. Mobile, Alabama, USA, March.

Scott, C.R. (1974) *Soil mechanics and foundations.* Applied Science Publishers Ltd., London.

Singleton, S.S. (1969) 'Grout hole orientation'. American Society of Civil Engineers, Journal of the Soil Mechanics Foundation Division.

Symposium on Grouts and Drilling Muds in Engineering Practice (1963). Butterworths, London.

Whittaker, W.H. (1980) 'Grouting machinery'. *Plant Engineering* (USA), **24** 2, 24th Jan.

CHAPTER EIGHT

Piling methods

Introduction

Many building projects and virtually all civil engineering construction works require the use of piling equipment during the ground engineering phase. Piles must be installed to support foundations, sheet pile walls are necessary to secure excavations, coffer-dams are needed to provide a dry site for river and coastal works, etc.

Usually the piles are driven into position with a pile hammer, except for those types of pile which are formed in situ in the ground. During pile-driving both pile and hammer must be temporarily held in place either from a crane jib, pile frame or from leaders. Also, piles for temporary works must often be removed after completion of the work, and a pile extractor is therefore required. All these aspects will be considered in turn in this chapter.

Pile hammers

Types of hammer

(i) Drop
(ii) Single-acting steam, compressed-air, or diesel
(iii) Double-acting steam, compressed-air, or diesel
impact hammers
(iv) Vibratory
(v) Hydraulic.

Drop Hammer

Simple hand-driven hammers (Fig. 8.1) have been used to drive light timber piles to shallow depths in soft earth, but today this method would seldom be selected. The more common choice is the power-assisted drop hammer (Fig. 8.2). The hammer weight is hung from a rope or cable running over a pulley to a rope drum. The whole

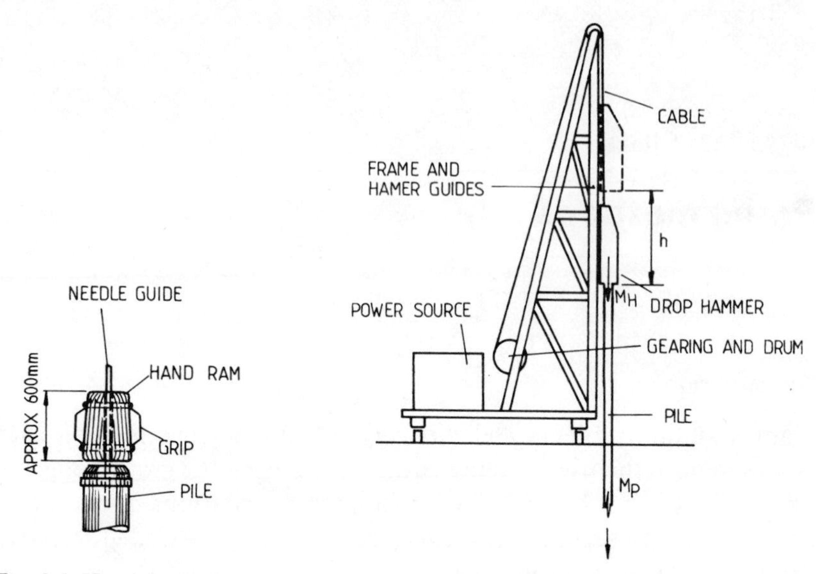

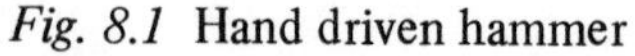

Fig. 8.1 Hand driven hammer

Fig. 8.2 Drop hammer

arrangement is supported on a strong frame or leader. The hammer is released by a manual trip and drops under free fall. It is raised by winching powered by a diesel or electric motor and suitable reduction gearing. About 10 – 20% of the potential energy is lost in the drag of the cable. This method is fairly slow, but is simple, requires little maintenance or specialist operators and is commonly used for all types of pile-driving.

Theoretical principles of the free fall hammer

Blow efficiency (*Fig. 8.2*)

The impact energy delivered by the blow is equal to the potential energy of the hammerhead, i.e.

$$E_i = M_H \times h \times g \qquad (8.1)$$

where E_i = impact energy per blow
M_H = mass of the hammerhead or ram
h = height of drop
g = force of gravity

Also, according to the principles of impact momentum the sum of momentums of hammerhead and pile before impact is equal to the combined momentum of pile and hammerhead after impact.

Thus

$$V(M_H + W_p) = M_H V_H + M_p V_p \tag{8.2}$$

where M_H = mass of the hammerhead
M_p = mass of the pile
V_H = velocity of hammerhead before impact
V_p = velocity of pile before impact
V = common velocity of hammerhead and pile immediately after impact

But $$V_H = \sqrt{2gh} \text{ and } V_p = 0$$

Thus $$V = \frac{M_H \times \sqrt{2gh}}{(M_H + M_p)} \tag{8.3}$$

The kinetic energy transmitted to the pile is E_k and

$$E_k = \tfrac{1}{2}(M_H + M_p) \times V^2 \tag{8.4}$$

by substituting V from (8.3) into (8.4)

$$E_k = \frac{M_H^2 \times g \times h}{M_H + M_p} \tag{8.5}$$

The efficiency of the blow η is

$$\eta = \frac{\text{transmitted energy } E_k}{\text{input energy } E_i} \tag{8.5a}$$

Thus

$$\eta = \frac{M_H^2 \times g \times h}{(M_H + M_p)} \times \frac{1}{M_H \times h \times g}$$

$$= \frac{M_H}{M_H + M_P} \text{ or } \frac{W_H}{W_H + W_p} \tag{8.6}$$

Note: $W = M \times g$

Thus it can be seen that when the weight of the pile and hammerhead are equal, the blow efficiency is only 50%. Good practice therefore requires that for effective pile-driving the weight of the hammerhead should be at least equal to the weight of the pile. When the ratio of hammerhead weight to pile weight falls below $\frac{1}{3}$, the driving effectiveness is seriously impeded.

It can also be seen in equation (8.1) that the input energy is proportional to the product of hammerhead weight and height of fall and may be increased by changing one or both of these factors. The energy transmitted is affected as follows:-
The energy wasted is $E_i - E_k$

$$E_i - E_k = 0.5\, M_H V_H^2 - 0.5\, (M_H + M_p)\, V^2 \quad (8.7)$$

Substituting for V

$$E_i - E_k = 0.5\, M_H V_H^2 - 0.5\, (M_H + M_p) \frac{M_H^2 V_H^2}{(M_H + M_p)^2}$$

$$= \frac{M_H M_p}{2(M_H + M_p)} \times V_H^2 \quad (8.8)$$

$$= \frac{M_H M_p}{(M_H + M_p)} \times gh \quad (8.9)$$

Thus a heavy hammerhead with a low velocity produces a higher blow efficiency than a light hammer with high velocity.

Also, when driving piles which are easily shattered, e.g. concrete, it is sensible to choose a heavy hammer with a low drop, thus reducing the impact velocity for a given level of input energy.

Because the drop hammer can supply a very heavy blow, it is suited to piles with high end bearing or which must overcome high soil resistance.

Drop hammer data

Weight	250 - 3000 kg
Height of drop	Up to 3 m
Rate of blows	5 - 15 blows per minute

Single-acting steam or compressed-air hammer

Single-acting steam or compressed-air hammers drive piles in a similar manner to the drop hammer, but the hammer is raised by steam or compressed-air rather than by winching.

An example of a typical steam hammer is shown in Fig. 8.3. It consists of a part hollow piston rod and sliding cylinder. Steam or air is admitted into the piston rod through a valve, by means of a lever which is either manually operated to control the hammer speed, or set automatically to give a preselected rate. The cylinder is thus raised up the piston shaft as the air enters the chamber. The lower part of the piston rod, which is solid, passes through the base of the cylinder to rest on the pile head, thus providing a steadying effect to the pile and exerting a slight downward force. To induce the hammer blow the lever is released, thus shutting off the air inlet valve and opening the exhaust valve to cause the cylinder to fall onto the anvil at the base of the piston rod.

A safety valve is usually incorporated in the cylinder wall to prevent

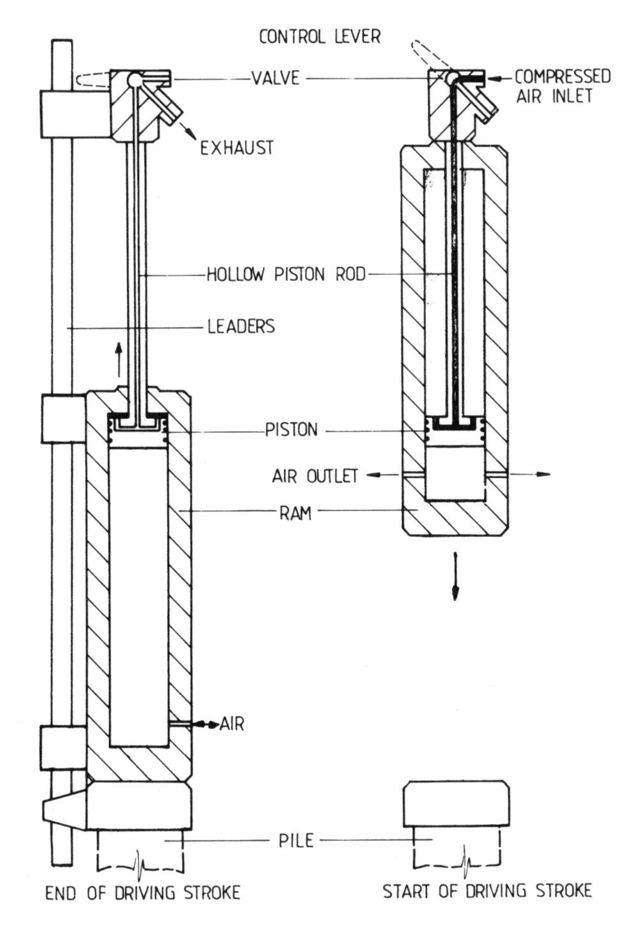

Fig. 8.3 Single-acting air or steam hammer

the hammer over-running on the upward stroke. Also, difficulties may occur in cold weather when the air or steam condenses on the inner wall of the cylinder and feed pipes, thus reducing the efficiency of the unit. The top of the piston rod is usually attached by rope to a winching mechanism, for raising and lowering the hammer when changing piles. The whole unit must be attached to leaders or a pile frame on a sliding guide, to provide directional control during the driving phase.

A higher striking rate is achieved compared with the drop hammer, but like the latter, a heavy blow (i.e. high impact energy) is delivered and it is thus not suited to driving light, thin-walled or sheet piles. Typical data are given in Table 8.1.

It can be seen from the data in Table 8.1 that an appropriate ram

Table 8.1 Single-acting hammer data

Overall height (m)	Ram weight (kg)	Stroke or drop height (m)	Impact energy per blow (kN/m)	Total operating weight (kg)	Blows per minute	Compressed-air consumption (m^3/min of free air)	Steam consumption per hour (kg)
4.2	1000	1.25	12.50	1400	35 - 45	15	400
4.2	1500	1.25	18.75	1900	35 - 45	18	500
4.5	2000	1.25	25.00	2500	35 - 45	20	600
4.5	2500	1.25	31.25	3000	35	22.5	700
4.6	3000	1.25	37.50	3700	35	25.5	770
4.6	4000	1.25	50.00	4800	25 - 30	28.5	910
5.0	5000	1.35	67.50	5900	25 - 30	37.0	1090
5.1	6000	1.35	81.00	7000	20 - 25	40.0	1360
5.1	8000	1.35	108.00	9000	20 - 25	42.0	1360
5.1	10000	1.35	135.00	11000	20	42.0	1360
5.1	12000	1.35	162.00	13800	20	51.0	1360

Note: (i) The blow rate can be increased to about 50 per minute on some models.
(ii) Single-acting hammers are available up to 140 000 kg ram weight for special duties.
(iii) A comparable range of differential-acting hammers are still manufactured. These hammers use the steam or air pressure acting on the piston in addition to gravity. Also, a higher rate of blows is obtained compared to the single-acting version.

weight may be selected to match a given weight of pile as described for the drop hammer. Indeed, for high soil resistances, hammers are currently available exceeding 12 tonnes weight.

Single-acting diesel hammer

In recent years the diesel hammer has proved to be a successful and in many cases a better alternative to the drop, single-acting steam or compressed-air hammer, because the available input energy per blow is roughly doubled for a comparable weight of ram. The action is similar to that of the air steam hammer, but is started by first raising the ram, which is automatically tripped to fall at the top of the stroke (Fig. 8.4(a)). As the ram falls, fuel is injected onto the impact block, and after passing the exhaust ports the air trapped in the cylinder is compressed. The impact of the ram on the impact block delivers energy to the pile but also causes combustion of the highly compressed air/fuel mixture, thus imparting further energy to the pile (Fig. 8.4(b)). The explosion causes the piston to move upwards and the gases are expelled

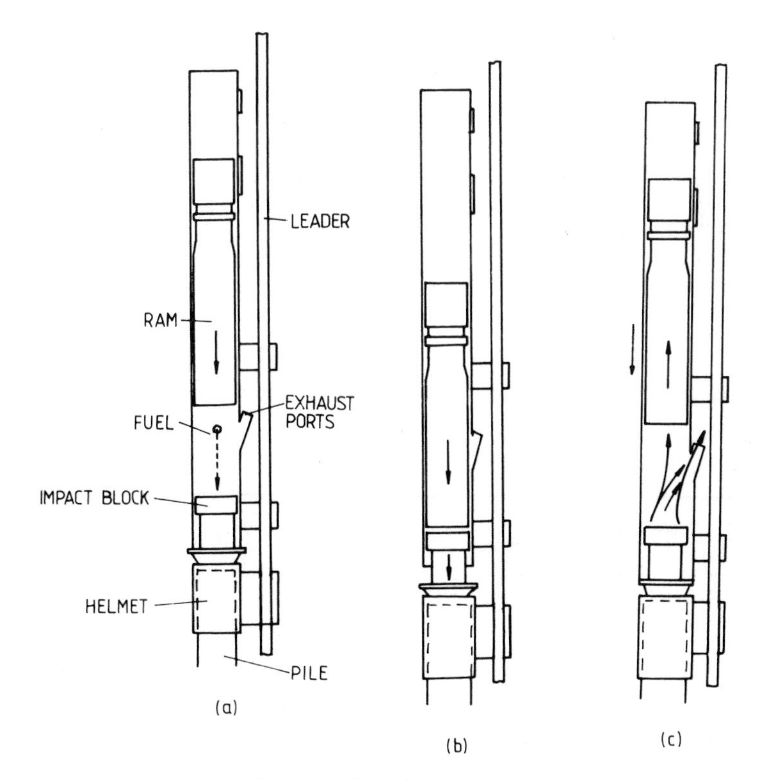

Fig. 8.4 Diesel hammer

through the exhaust ports (Fig. 8.4(c)). The cycle will continue until the fuel supply is cut off. Typical data are given in Table 8.2.

A diesel hammer ram weighs less than the equivalent single-acting steam or compressed-air hammer and does not require a steam boiler or compressed-air supply. However, starting may be difficult in soft ground, when the impact of the falling ram may not be sufficient to atomise the fuel. There is also a tendency for the stroke to increase to the maximum with increasing penetration resistance. While this can be an advantage, care should be taken not to damage the pile head.

Like other single-acting hammers the diesel hammer is most suitable where a very heavy blow (i.e. high impact energy) is required. The weight of the ram should be at least equal to the weight of the pile for efficient pile-driving.

Table 8.2 Single-acting diesel hammer data

Ram weight (kg)	Stroke (m)	Impact energy per blow (kN/m)	Blows per minute	Total operating weight (kg)	Fuel consumption (1/h)	Overall height (m)
200	0.6 - 1.3	1.5 - 2.5	60 - 70	350	0.5	2.0
400	0.8 - 1.6	2.2 - 5.0	50 - 60	650	0.8	2.4
500	1.2 - 2.0	12.5	40 - 60	1250	3.5	3.8
1000	1.2 - 2.0	30.0	40 - 60	2500	6.5	4.2
1500	1.2 - 2.0	37.5	40 - 60	3000	6.5	4.5
2250	1.2 - 2.0	33.5 - 67.0	35 - 50	5000	7.0	5.0
3000	1.2 - 2.0	45.5 - 91.0	35 - 50	6000	8.0	5.2
3500	1.2 - 2.0	55.0 - 110.0	35 - 50	8000	11.5	5.2
4500	1.2 - 2.0	70.0 - 140.0	35 - 50	9000	16.5	5.2
5500	1.2 - 2.0	90.0 - 180.0	35 - 45	12000	20.0	5.9

Double-acting steam or compressed-air hammer (*Fig. 8.5*)

The double-acting hammer usually employs air or steam, which is admitted to the upper and lower cylinders alternately by means of a valve actuated by the piston. In this way both free fall impact and additional energy from the release of compressed-air into the upper cylinder are obtained on the downward stroke. By switching the air supply to the lower cylinder, the piston is raised and the air in the upper cylinder expelled, ready to repeat the cycle. The stroke of double-acting hammers is quite small, usually less than 0.3 m. They are designed with relatively light rams for pile-driving in loose soil and for light sheet piles and thin-walled piles, which offer relatively little resistance to penetration. Only sufficient impact energy is provided to

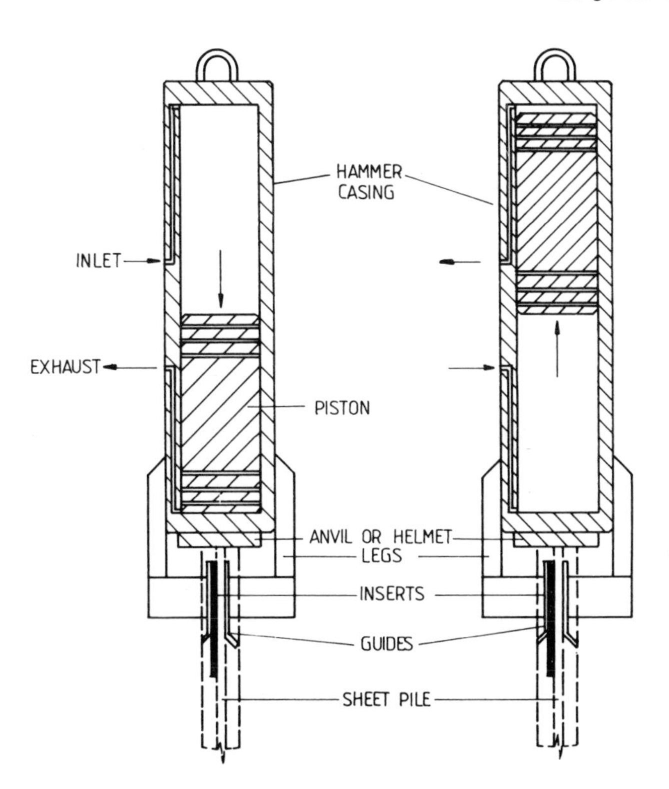

Fig. 8.5 Double-acting air or steam hammer

avoid undue damage to the top of the pile. It can be seen from Tables 8.1 and 8.3 that the largest double-acting air hammer is approximately equivalent in size and energy to medium range single-acting hammers. For heavier piles, double-acting diesel hammers are available with similar energy delivery characteristics to the single-acting versions.

Double-acting hammers run at considerably greater speeds than single-acting or diesel hammers, thus ensuring that the pile is maintained in almost continuous motion, which is particularly advantageous in overcoming static friction in granular soils. Double-acting hammers, however, are less effective in cohesive soils such as clay, where a heavier free fall hammer may be more suitable. Also, because of the rapid blow delivery rate, concrete piles may suffer head damage by shattering. Finally, double-acting hammers can be operated successfully under water, when fitted with an exhaust extension.

Currently, developments are taking place with hydraulic impact hammers; however, experience with this method is limited, except to say that noise levels are significantly reduced.

Theoretical energy delivery of double-acting hammer

$$E_i = [W_H + (A_p \times p)] \times h \qquad (8.8)$$

where E_i = impact energy per blow
W_H = ram weight
A_p = surface area of piston
p = average pressure in cylinder
h = stroke

Energy delivered per minute = $E_i \times n$, where n is the number of blows per minute.
The input energy ($A_p \times p \times h$) indicated in eqn. 8.8, simply increases E_i compared to the free fall condition.
Thus from equation (8.5a) it can be seen that the blow efficiency (γ) may be increased by using a heavy hammer (M_H) relative to the pile weight (M_p).

Table 8.3 Double-acting hammer data

Weight of ram (kg)	Input energy per blow (kN.m)	Total operating* weight (kg)	Overall† height (m)	Air consumption (m^3/min)	Steam consumption (kg/h)	Blows per min
10	0.15	65	0.7	2	140	500
20	0.22	150	0.8	2	140	500
30	0.50	300	1.5	3	200	400
90	1.40	700	1.5	7	270	300
180	3.50	1300	1.6	11	360	275
260	5.70	2300	1.8	13	480	225
720	12.00	3200	2.5	17	610	145
1400	18.00	5000	2.8	21	680	105
2300	26.00	6400	3.4	25	950	95

* Including legs for driving sheet piling
† Excluding legs
Air pressure 6.5 bar (90 p.s.i.)
Steam pressure 10 bar (140 p.s.i.)

Vibratory pile driver

In recent years there has been a demand for a quiet method of driving piles, for example near hospitals and in residential areas. For sheet piles and other types of steel pile a vibratory technique has proved successful in granular type soils.

The vibratory pile-driver (Fig. 8.6) consists of an electric or hydraulic

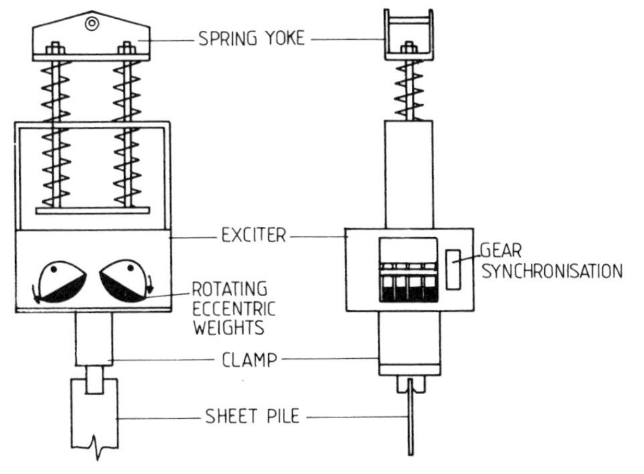

Fig. 8.6 Vibratory pile-driver

motor coupled to two eccentrically and separately mounted cams which rotate in synchronised opposition to each other. The whole unit is housed in a steel casing and for operation is suspended from the lifting rope of a crane. A suspension bracket, spring-mounted to the pile casing, eliminates upward vibration into the rope.

The vibrator may be attached to the pile by means of remotely controlled, hydraulically-operated grips.

The action on the pile is produced by the rotation in unison of the two eccentric cams. During each cycle the centrifugal forces act downwards at 0° and 360°, and upwards at 180° from the vertical, while at all other positions the opposing horizontal components of their respective forces are cancelled. The pile is thus momentarily shaken up and down, thereby reducing the static friction between the soil and pile. Thus when the end bearing area of the pile is small, the downward force produced by the self-weight of the driver and pile causes the pile to sink into the soil.

The effectiveness of the vibrator can be improved by regulating the frequency (i.e. altering the rotation speed of the cams) to suit the particular soil conditions and to avoid resonance in nearby structures. The amplitude of vibration may also be adjusted by altering the mass of the eccentric weights.

The vibratory pile-driver is most efficient in water-bearing sands, and is effective in most loose to medium sands and gravels. Some difficulty may be encountered in very dense granular material, where the downward dynamic force may not be sufficient to produce adequate penetration of the pile. The method is less effective in cohesive soil.

However, the Bodine Sonic pile-driver, and more recently the hammer developed by Hawker Siddeley Ltd., which operate at frequencies up to 150 hertz (9000 cpm) and thereby put the pile at its resonant frequency, have been used in fairly stiff clay. The result is a minute and momentary alternate increase and decrease in the cross-section thickness of the pile thus breaking the cohesion between pile and soil. The downward driving force is simply supplied by the pile-driver and self-weight of the pile.

The main advantage of the vibratory pile-driver is the relative lack of noise and absence of exhaust fumes. In addition, the pile head suffers little damage compared to hammer methods. The method can also be used to extract piles, by simply providing an upwards pull of perhaps 15 – 25 tonnes from the crane rope.

Table 8.4 Vibratory pile data

Max. dynamic force (kN)	Suspended weight (kg)	Power (kW)	Max. frequency (Hz)	Max. crane pull for use as an extractor (tonnes)	Duties
200	1200	50	35	7	Light duty sheet piles etc.
400	2000	100	25	20	Light duty sheet piles, open-ended tubes up to 500 mm diameter.
650	4000	175	25	30	Medium duty sheet piles, columns, open-ended tubes up to 6 tonnes wt.
1400	10000	350	25	40	Heavy duty sheet piles, beams, open-ended tubes up to 20 tonnes wt.
2000	16000	550	20	80	Heavy columns, large diameter tubes, generally heavy sections.
4000	30000	800	20	80	Very heavy sections

Note: Vibratory pile-drivers may be powered electrically from the main or a generating set (400 – 450 V), or hydraulically from a diesel engine.

Hydraulic sheet pile-driver

While the vibratory pile-driver is suitable for 'silent' driving most types of pile in sands, a quiet pile-driver is often required for cohesive soils.

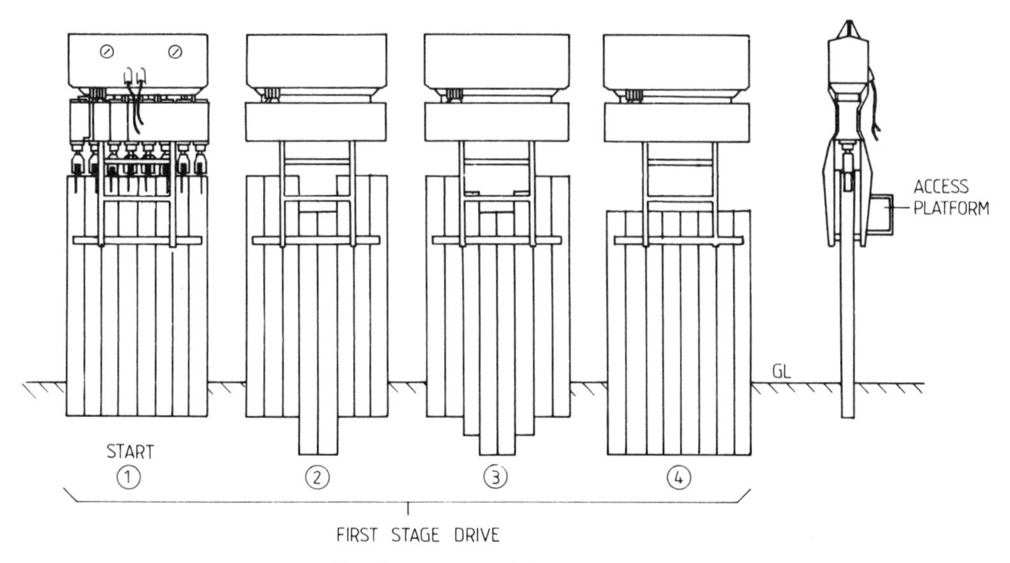

Fig. 8.7 Taywood Pilemaster

In the early 1960s Taylor Woodrow Construction Ltd. set about solving the problem for sheet piling, and developed the Taywood Pilemaster (Fig. 8.7). The Pilemaster consists of an electric motor, hydraulic pumps, fuel tank, etc. mounted on a crosshead fabricated from steel plate. Eight hydraulic jacks are attached to the base flange of the crosshead. Each jack works on the principle as shown in Fig. 8.8, where oil at a pressure of approximately 62 N/mm^2 (620 bar or 9000 p.s.i.) is introduced into the cylinder through a valve, thereby raising or forcing the piston downwards for pile-driving. Each jack is connected to a friction plate by a fork-shaped high tensile forged steel connector (Fig. 8.9). The friction plate is then bolted to the pile. The connector head may be rotated to accommodate attachment to most shapes of pile and the hemispherical seating permits up to 3° misalignment between the pile-driver and piles.

Pile-driving method

The Pilemaster drives in panels of seven or eight piles, and bends or junctions must usually be overcome with specially shaped piles. Piles are normally pitched in a temporary frame to give initial support (see Fig. 8.39). The pile-driver itself weighs about 10 -12 tonnes and is hung from a crane jib. Guide legs maintain the machine in alignment with the pile, and driving usually starts with the centre piles, working out to the end piles in the panel (see Fig. 8.7). When all the piles have been driven the full stroke of the rams, the Pilemaster is lowered to the new level and the procedure repeated.

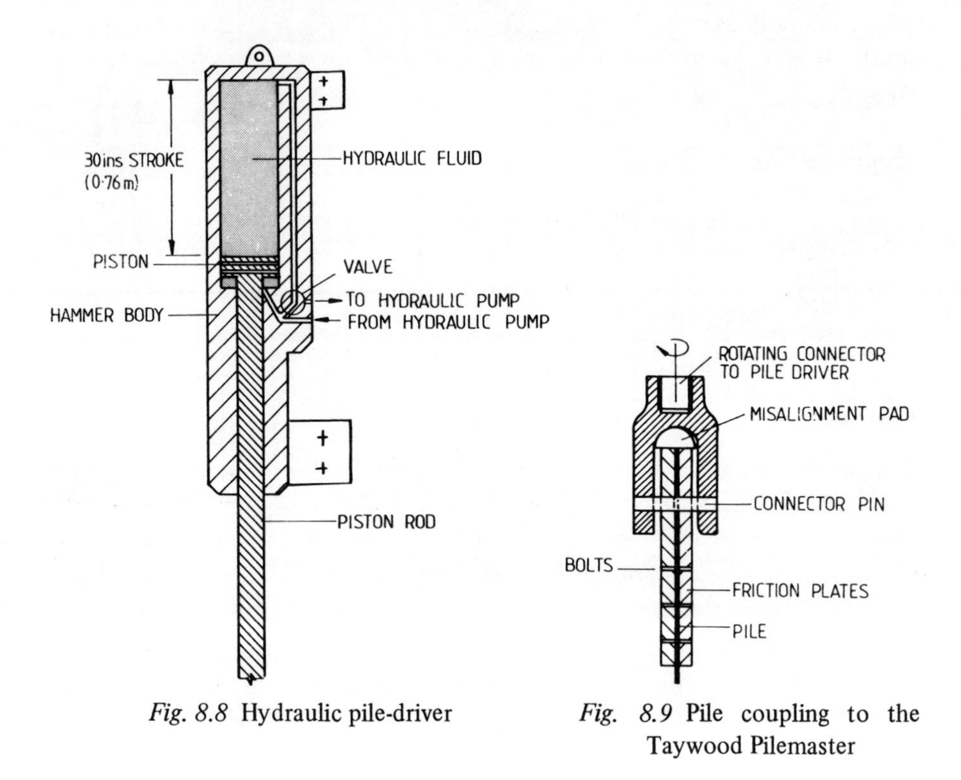

Fig. 8.8 Hydraulic pile-driver

Fig. 8.9 Pile coupling to the Taywood Pilemaster

The initial reaction for driving the first pair of piles is obtained from the weight of Pilemaster and the weight of the panel of piles attached to it. The friction and restraint provided by the first pair are then used to give additional restraint to drive the next pair. A progressive accumulation of frictional restraint is therefore generated as the piles penetrate deeper into the soil, until the restraint is almost entirely frictional.

Support from the crane may be dispensed with when the panel becomes self standing, thereby freeing lifting capacity for pitching the adjacent panel. If the driving is to continue to ground level, piles should be initially pitched in a shallow trench to facilitate removal of the friction plate.

The Pilemaster has proved successful in loose to medium dense fine granular sands and in cohesive soils such as clay. Coarse granular material offers too high a toe resistance for efficient pile-driving.

The load supplied to each jack is about 250 tonnes, which enables driving depths of up to 20 m in favourable soils.

The driving speed varies according to the driving resistance encountered, e.g. 0.5 m/min at high pressure, to 1 m/min at low pressure operation. The Pilemaster can be operated with most types of sheet pile and with suitable support rakes, up to 1 in 5 may be obtained. In addition the machine may be used to extract sheet piles from all types of soil.

Pilemaster data

Pile weight	10 - 12 tonnes
Power	30 kW
Voltage	400 - 450 volts
Driving force per ram	200 - 250 tonnes
Length of pile	15 - 20 m

Pile extractors

The most common type of extractor operates on the principle of the inverted pile hammer using steam or compressed-air. In the example shown in Fig. 8.10, air is admitted to the upper cylinder via a valve causing upward movement of the casing until the base of the cylinder meets the underside of the ram, thereby transferring an upward blow to

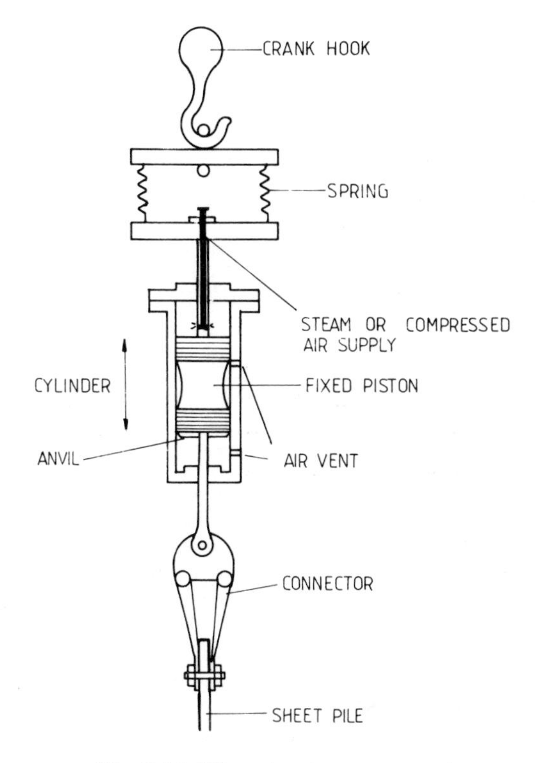

Fig. 8.10 Pile extractor

the pile. The whole unit is hung from a crane or frame via a spring mechanism to prevent vibrations being transferred to the lifting rope.

Pile extractors are available for removing light trench sheeting up to timber and heavy sheet and H piles.

Table 8.5 Pile extractor data

	Input energy per blow (kN/m)	Weight of ram (kg)	Air* consumption (m^3/min)	Steam† consumption (kg/h)	Blows per min
Light duty	0.35	10	3.5	–	1000
	0.75	20	5.5	–	800
	1.50	40	10.0	–	700
	3.00	80	10.0	–	400
Heavy duty	8.00	500	10.0	350	150
	11.00	750	10.0	500	140
	16.00	1750	13.0	600	120
	30.00	3500	16.0	800	90

* 6.5 bar (90 p.s.i.)
† 10 bar (140 p.s.i.)

Other methods of pile extraction include vibration and hydraulic jacks.

Pile-driving assisted by water jet

Work done in Japan has demonstrated the application of a water jet in assisting the driving of sheet piles through mudstone (Fig. 8.11) with a vibratory hammer. Water is pumped through a 25 mm diameter pipe

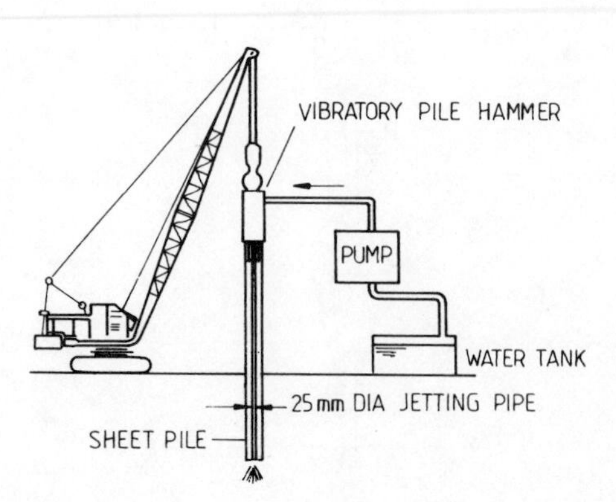

Fig. 8.11 Jetting assistance for sheet pile-driving

attached to the side of the pile with a water jet, giving a 30° angle of spread. The method can be used to break up the hard material ahead of the tip of the sheet pile, and it is claimed that the rate of driving may be doubled compared with using the vibro hammer by itself. Water flow of 5 – 15 m^3/h is necessary.

Noise

Noise is a particularly unpleasant feature of pile hammers and manufacturers have given some attention to noise reduction, but as yet with only moderate success. The principal method of reducing noise is totally to enclose the hammer, leaders and pile in a large soundproof box, the leaders being used to form the box framework. Clearly, this method is cumbersome, and cannot be easily adapted for sheet piling. More satisfactory methods use vibration or hydraulic methods, but these are generally more expensive.

Pile helmets and dollies

Helmets

A pile helmet is required to distribute the blow from the hammer evenly to the head of the pile, to cushion the blow and to protect the pile head itself.

Helmets with a relatively thin base can be used with solid piles (Fig. 8.12) because the load is transferred across the whole pile head. However, pipes, sheet piling and other thin-walled piles require a much thicker base (Fig. 8.13) to ensure that the impact energy is directly transferred into the pile wall. The helmet should have about 10 mm clearance around the pile (Fig. 8.14) to avoid damage to the pile head.

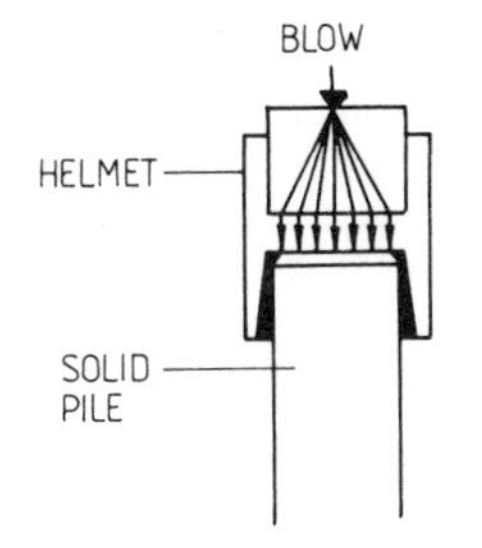

Fig. 8.12 Pile helmet for solid piles

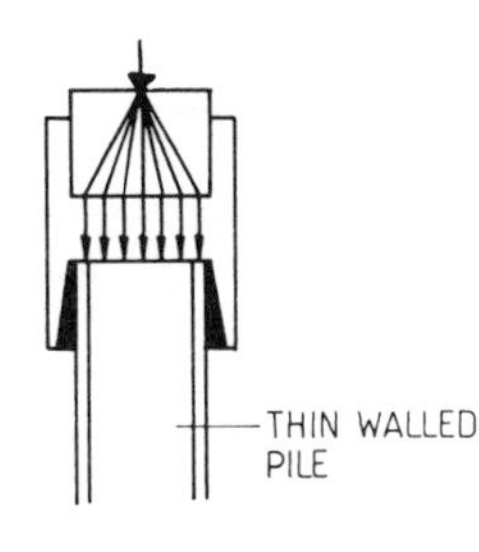

Fig. 8.13 Pile helmet for thin-walled piles

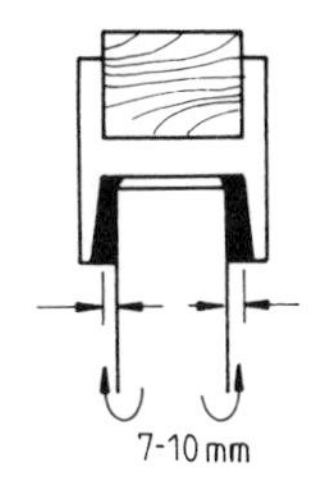

Fig. 8.14 Dolly used to cushion the blow

Dollies

The dolly is placed in a recess in the helmet and acts to cushion the blow (Fig. 8.14). The effect is to extend the impact time by storing the impact energy in the dolly, as indicated in Fig. 8.15. By selecting the appropriate material, a better transfer of energy from hammer to pile can be obtained in different ground conditions.

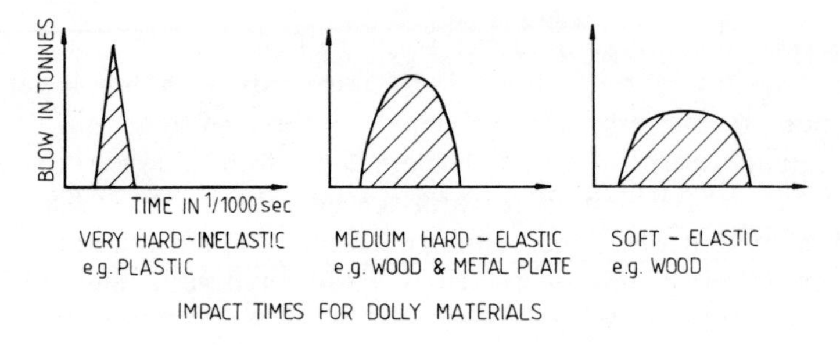

Fig. 8.15 Impact times for dolly materials (Delmag)

For medium to heavy penetration resistances, a plastic or resin-bonded fabric dolly with a steel plate is preferred (Fig. 8.16). Alternatively, a wooden dolly made from oak or similar material capped with a steel plate can be used; this combination tends to give more recoil (Fig. 8.17). For light to medium driving, a simple wooden dolly of beech or elm is adequate (Fig. 8.18).

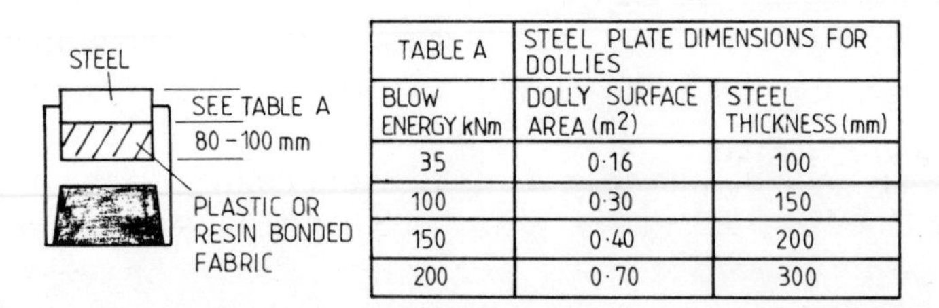

TABLE A	STEEL PLATE DIMENSIONS FOR DOLLIES	
BLOW ENERGY kNm	DOLLY SURFACE AREA (m2)	STEEL THICKNESS (mm)
35	0·16	100
100	0·30	150
150	0·40	200
200	0·70	300

Fig. 8.16 Composite dolly comprising steel plate and resin bonded fabric

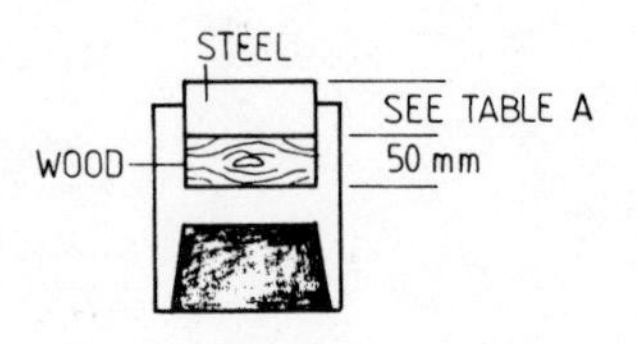

Fig. 8.17 Composite dolly comprising steel plate and wood

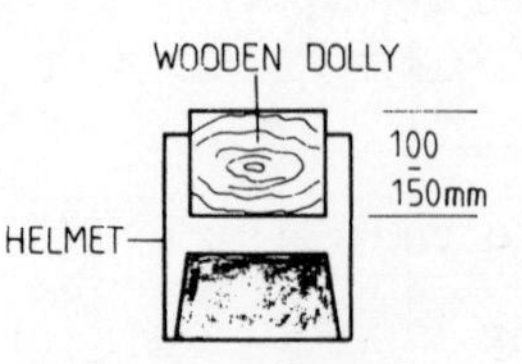

Fig. 8.18 Wooden dolly

Plastic dollies may last for several hundred piles in moderate driving conditions, but unfortunately they produce a much harsher effect on the pile head, which may cause damage. Wooden dollies, however, tend to disintegrate quickly (i.e. 2 – 5 piles) if used in heavy driving conditions, and if not replaced when smoking and burning is observed, much of the impact energy will be absorbed, thus reducing the effectiveness of the blow.

Packing (*Fig. 8.19*)

Concrete piles and other materials are likely to suffer damage from the force of impact and require a cushion placed between the pile head and underside of the helmet. In this way, peak impact forces are avoided and the forces are transmitted uniformly into the pile.

Packing plates (Fig. 8.20) of 25 – 30 mm minimum thickness made of soft plywood are suitable in soft ground conditions; alternatives are paper bags filled with sawdust. For hard driving conditions asbestos fibre has proved adequate.

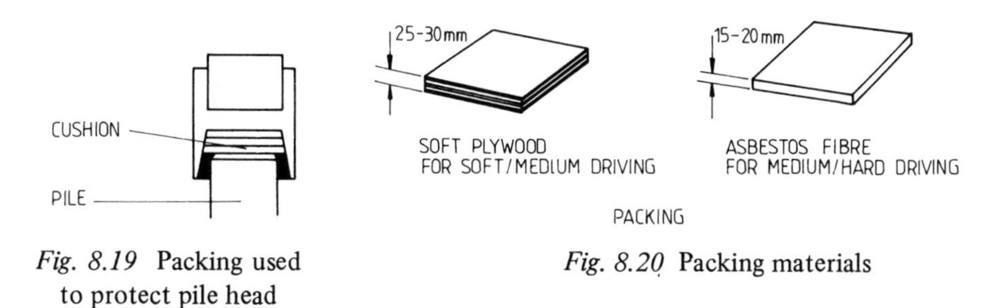

Fig. 8.19 Packing used to protect pile head

Fig. 8.20 Packing materials

Leaders and pile frames

In order to drive piles accurately, it is necessary to pitch and hold the pile in position, support the hammer and guide both the pile and hammer along the required drive angle. The choice of method varies according to weight of hammer and pile, ground conditions, length of pile, and manoeuvrability demanded. Depending upon these factors pile frames, hanging leaders or rope-suspended leaders may be selected.

Hanging leaders (*Fig. 8.21*)

Hanging leaders consist of a sturdy lattice-framed mast attached to the jib of a crane. The rake of the mast may be adjusted to suit forward, backward and lateral piling angles, thereby allowing piling to

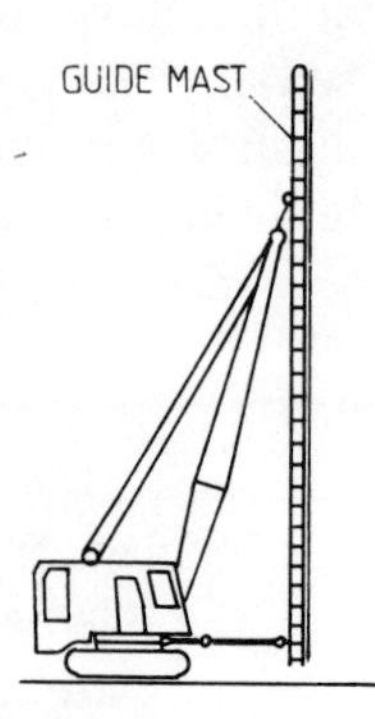

Fig. 8.21 Hanging leaders

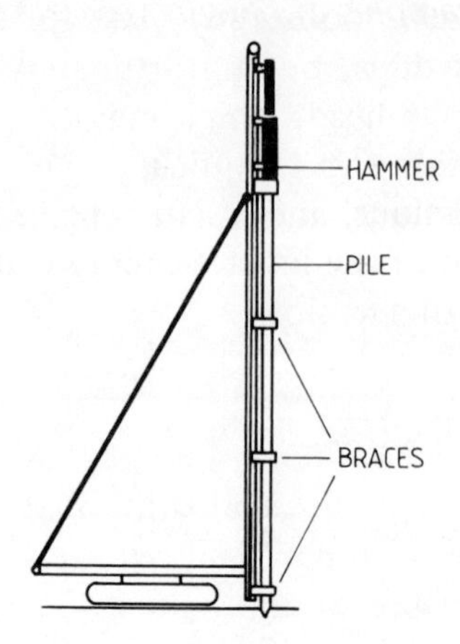

Fig. 8.22 Use of guides to secure pile and hammer

take place on uneven or sloping ground. The mast is moved into the desired position by means of hydraulically-controlled rams attached between the base of the jib and mast. The machine requires separate winches to lift and pitch the pile and to hold the hammer. The hammer is usually attached to guides on the leader as shown in Fig. 8.22. Guides are also necessary to control the direction of the pile during driving (Fig. 8.23) especially when using slender piles, when flexing may occur. Wherever possible, it is desirable to attach the helmet to the leader guides, both to provide directional restraint and to avoid losing the helmet during recoil in hard driving conditions.

The crawler-mounted hanging leader is particularly useful for driving on uneven or poor ground and in awkward positions. The weight of the hammer, pile and leaders must be within the lifting capacity of the crane, thereby limiting choice of the method to medium size duties.

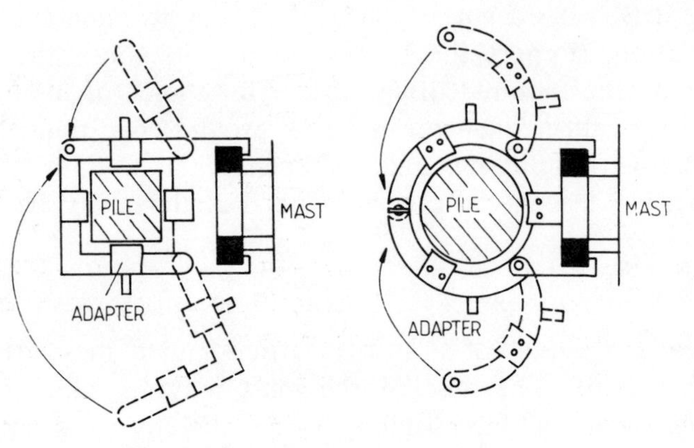

Fig. 8.23 Guide clamp details (Delmag)

Examples of hanging leaders data

	Small rig	*Large rig*
Height (H)	12 m	30 m
Hammer	800 kg	13000 kg
Pile	700 kg	15000 kg
Leader	650 kg	15000 kg
Forward and rear rake	1:3	1:3
Lateral rake	1:3	1:3

Pile frame (*Fig. 8.24*)

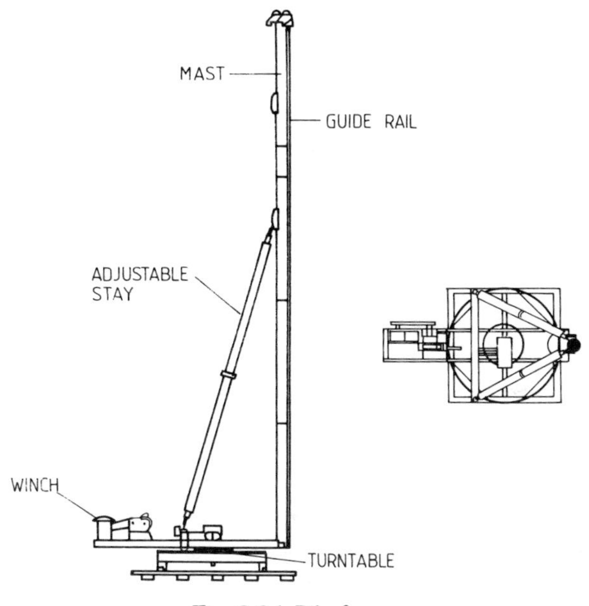

Fig. 8.24 Pile frame

When piles are either too long or too heavy for hanging leaders, or where it is convenient to mount the pile rig on rails, the pile frame is often the most suitable method to produce accurate directional control. The method is similar to that for hanging leaders and small economical rigs or large frames are available.

Modern frames consist of a leader supported by two adjustable struts to allow forward and backward raking. The mast can be extended at the bottom to facilitate pile-driving over water.

The base frame rests on four swivel steel castors so that the rigs may be moved around on hard ground or pads. However, for installing rows of piles, the manoeuvrability of the rig can be improved by placing the frame on a turntable to permit 360° turning.

The rig is equipped with winches and the whole unit can be erected in about ½ day by four men.

Examples of pile frame data

	Small frame	*Large frame*
Height	12 m	50 m
Weight of hammer, plus pile	6000 kg	40000 kg

Rope-suspended leaders (*Fig. 8.25*)

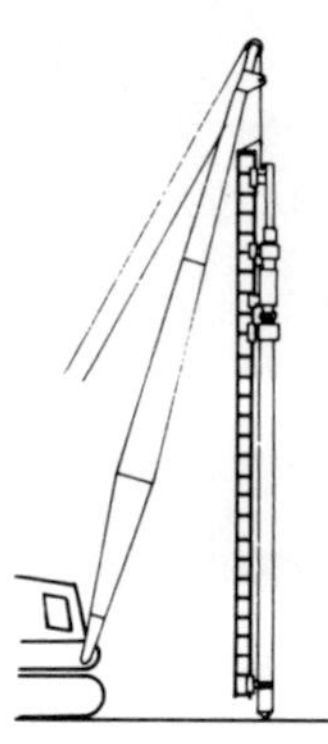

Fig. 8.25 Rope suspended leaders

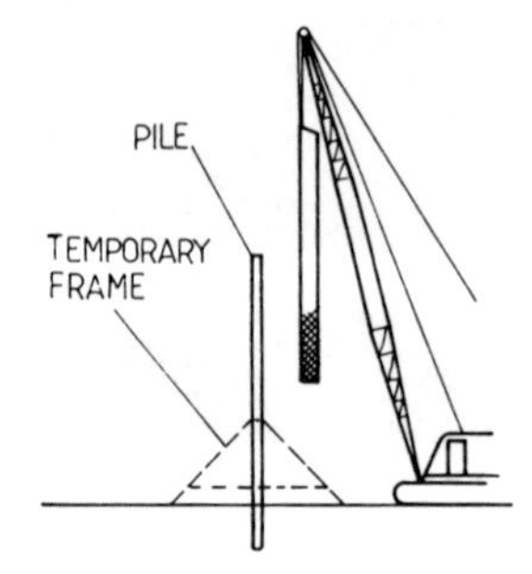

Fig. 8.26 Temporary frame to hold a pile

Piles supported in temporary framework (Fig. 8.26) are usually driven from rope-suspended leaders. Indeed, such leaders may be used to pitch and drive piles without the temporary support (Fig. 8.27) but there is then more danger that the pile will veer off the required direction if obstructions are encountered. Rope-suspended leaders are suitable for driving on a forward or lateral rake (Fig. 8.28).

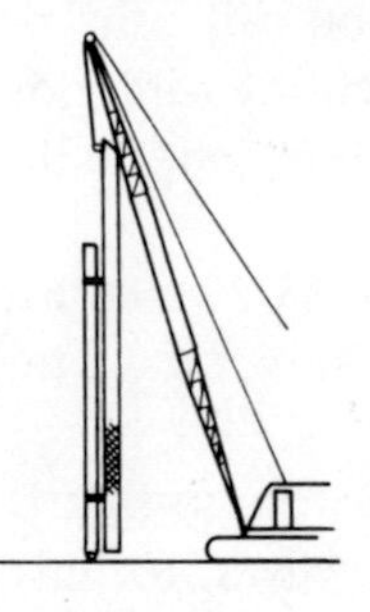

Fig. 8.27 Pile supported by rope suspended leader

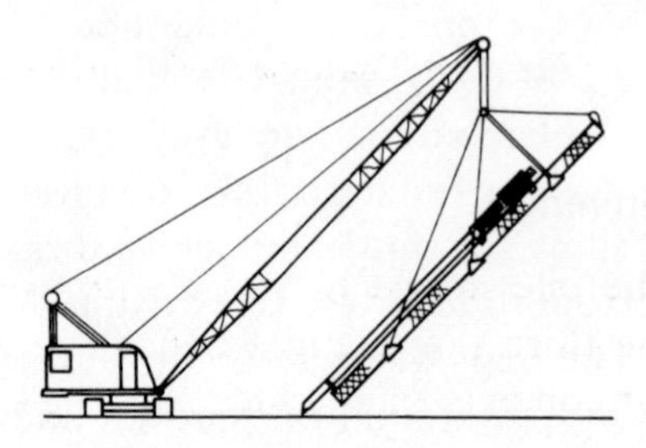

Fig. 8.28 Pile driven on the rake with rope suspended leaders

Example of rope-suspended leaders data

	Small	*Large*
Height	12 m	30 m
Weight of hammer	1000 kg	10000 kg
Weight of leader	600 kg	7000 kg
Weight of pile	Varies	Varies

Like the hanging leader, its size is governed by the lifting capacity of the crane.

Handling piles

Dragging

A pile should never be dragged along the ground from the top of the leader as this may cause overturning of the rig. Always use the bottom pulling technique (Fig. 8.29).

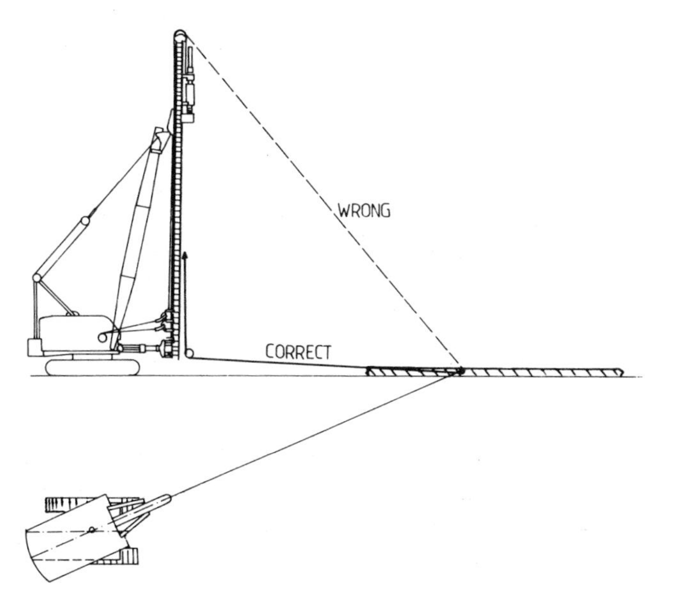

Fig. 8.29 Dragging a pile ready for positioning in the frame

Pitching

The pile should be raised into position carefully and lifted from about the third point (Fig. 8.30). These requirements are especially important for concrete piles which can easily crack.

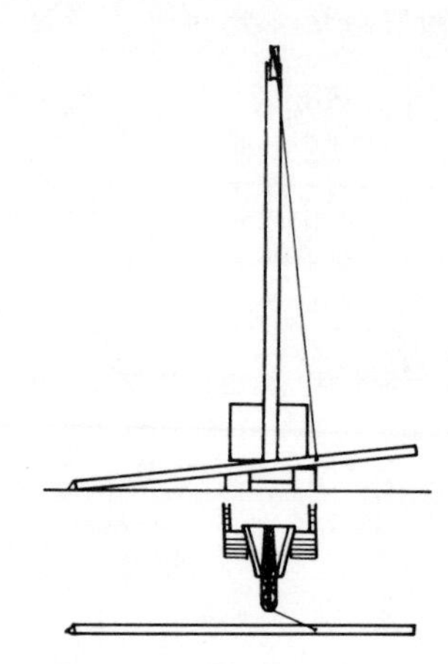

Fig. 8.30 Pitching a pile

Transporting

To avoid fracture or cracking, the pile should be lifted and carried at the points shown in Fig. 8.31.

If flexing is excessive, extra pick-up points can be provided from a temporary beam (Fig. 8.32).

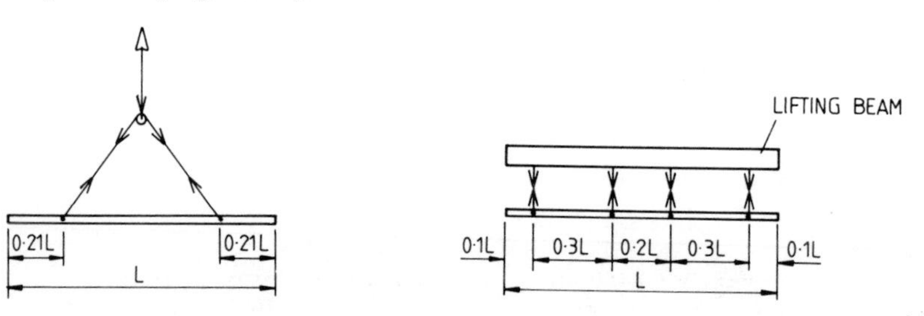

Fig. 8.31 Transporting a pile

Fig. 8:32 Transporting a long pile using a lifting beam

Sheet piles

Until fairly recently timber was almost universally used for shoring during excavation operations, but the escalation of prices for wood, coupled with the development of more rapid systems of erection, has forced engineers to turn to steel, and in particular to sheet piles. The modern sheet pile has several advantages, for instance:

(i) high strength is combined with light weight
(ii) the pile can be used many times over
(iii) the pile can be driven much more quickly than a timber pile

(iv) standard shapes and sizes are available to the designer
(v) steel may be used for either permanent or temporary works
(vi) piles may be interlocked for rigidity and reasonable watertightness
(vii) piles can be extended by welding
(viii) long piles can be driven with heavy pile hammers
(ix) piles are easy to handle and store on site
(x) a range of special piles can be made for interlocking corners, junctions, etc.
(xi) the pile is strong along its length in bending, allowing internal bracing inside coffer-dams to be replaced with ground anchors.

Basically sheet piles are manufactured U-shape, Z-shape, and straight (Fig. 8.33).

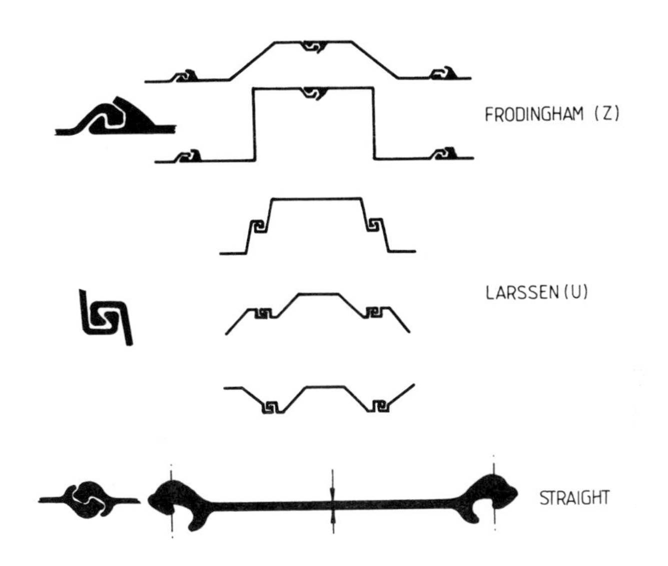

Fig. 8.33 Sheet pile types

Both Larssen U and Frodingham Z piles are equally suitable for most uses and, in general, contractors select the type most readily available. Applications include bulkheads, coffer-dams and retaining walls. Greater strength in bending can be obtained from strengthened piles e.g. composite construction (Fig. 8.34) or interlocking H piles. Straight web piles have little strength in bending and are mainly used for cellular coffer-dams, where interlocking piles are required to provide strength in tension. This type of structure is covered more fully in Chapter 9.

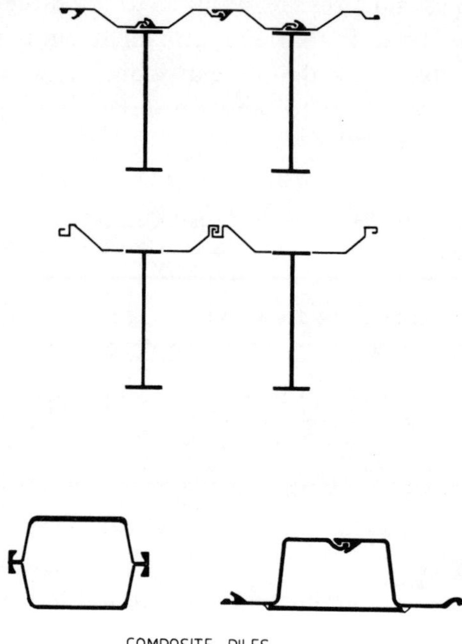

Fig. 8.34 Composite piles

Interlocks

The interlocks must be fairly loose to allow free sliding during driving. Consequently the joints cannot be made completely watertight. However, grease applied to the joint before driving helps to reduce friction and improve the driving, and also attracts fine particles into the joint.

Interlocks permit a deviation of only about 10° between adjacent piles, hence changes in wall direction generally require specially pre-bent piles to produce a 'watertight' structure.

Special piles

Special piles can be obtained for any angle up to 90°, (Fig. 8.35). Tee, cross and Y piles fabricated by riveting or welding sections together are also illustrated.

Trench sheeting

The U pile is available with very shallow sections for use as trench sheeting (Fig. 8.36). Applications include supports to the sides of

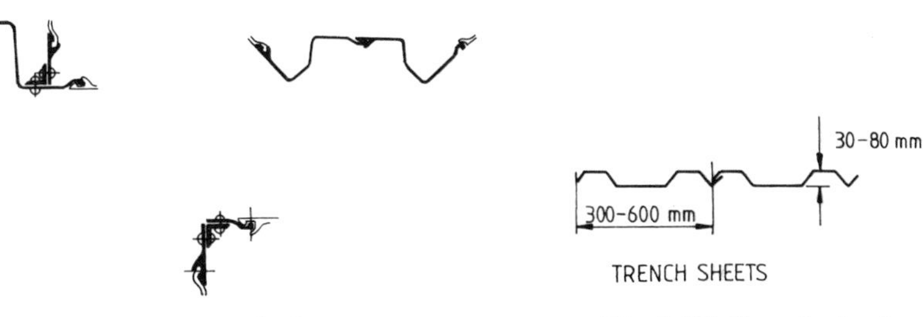

Fig. 8.35 Special piles

Fig. 8.36 Trench sheets

shallow trenches and excavations, temporary retaining walls along rivers and canals, etc. The piles are normally lap-jointed only, to aid speedy installation, but interlocking sections are available, and mild steel or high yield steel sections may be used.

Sheet Pile-driving

Pitching and guiding

Sheet piling is driven in panels of 6 to 10 pairs of piles in order to maintain accuracy in both the vertical and horizontal directions. Each pile is usually supplied with a hole drilled near the top to which a 'quick-release' shackle (Fig. 8.37) can be attached. A crane is then normally used to lift and pitch the pile.

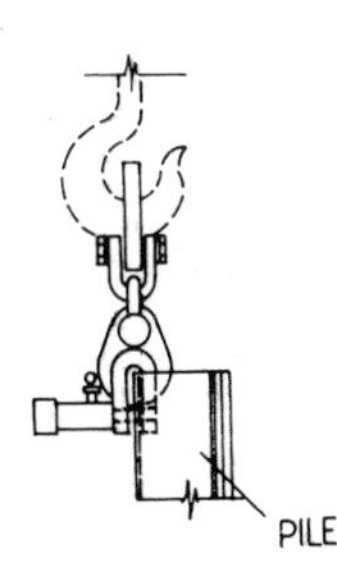

Fig. 8.37 Lifting bracket for sheet piles

Forming the initial interlock can be troublesome, especially in windy conditions, but this problem can largely be overcome by attaching the lightweight aluminium guide bracket developed by the Dawson Company (Fig. 8.38). A sturdy frame made of heavy timbers (Fig. 8.39) acts to guide the pile and to counter wind forces. Walings should be

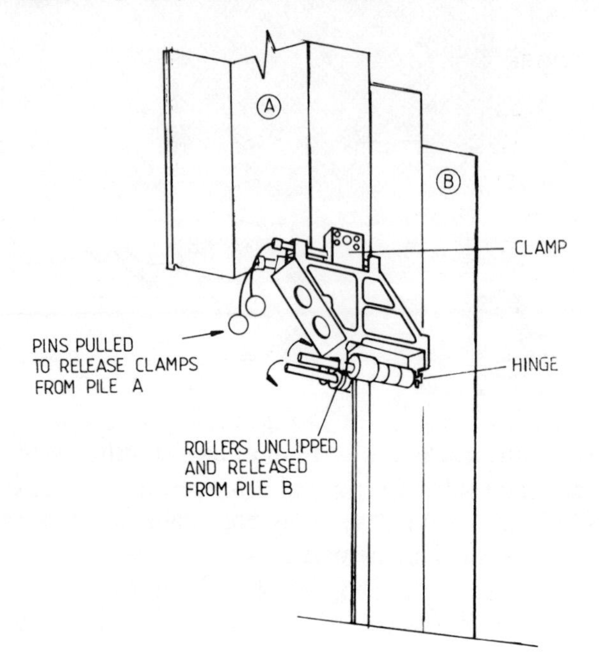

Fig. 8.38 The Dawson sheet pile guide bracket

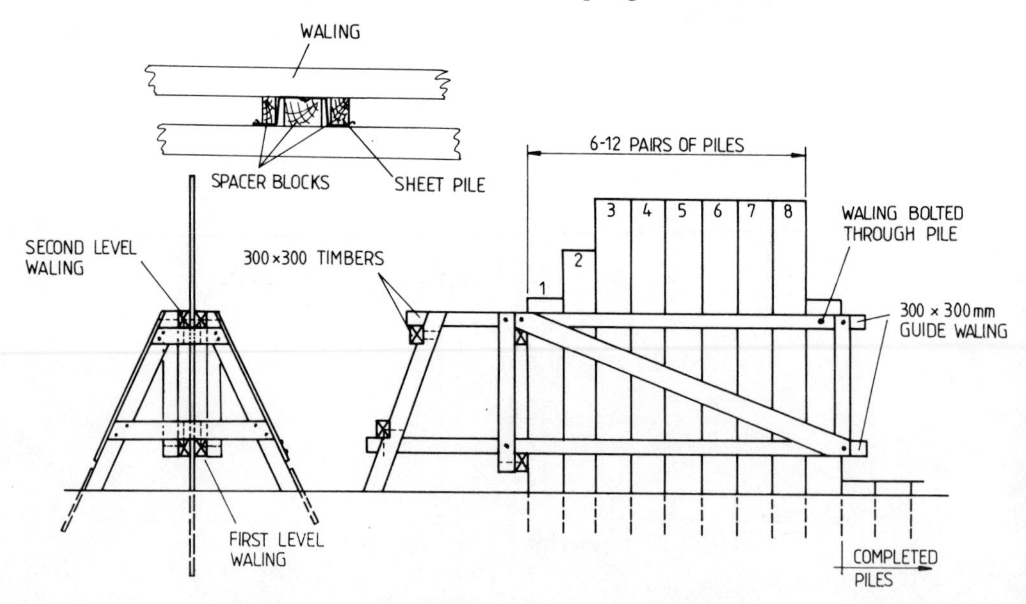

Fig. 8.39 Timber frame for sheet piling

provided at ground level with a second set as far away as practicable from the first to provide rigidity and directional restraint during driving.

Pile-driving procedure

The piles are usually pitched in pairs to form a panel and the first and last pairs are partially driven first, which helps to prevent creep, as there is some play in the pile interlocks. The remaining piles are then driven down to the level of the top waling. The hammer is hung from a crane boom and usually a leader is required when using a diesel or single-acting hammer. This can be avoided with the double-acting hammer, where integral guide legs (Fig. 8.5) keep it on the pile and aid vertical driving. Sheet piling is quickly damaged if the hammer is allowed to lean over.

A panel of piles is often sufficiently stable on reaching the top waling for pile-driving to continue to ground level with this waling removed. For more accurate work, however, either hanging leaders or even a pile frame should be used for the final stage of ramming. When long runs of piles are involved it may be necessary to have several sets of guide frames to allow the following panels to be pitched and driven in stages as demonstrated in Fig. 8.40.

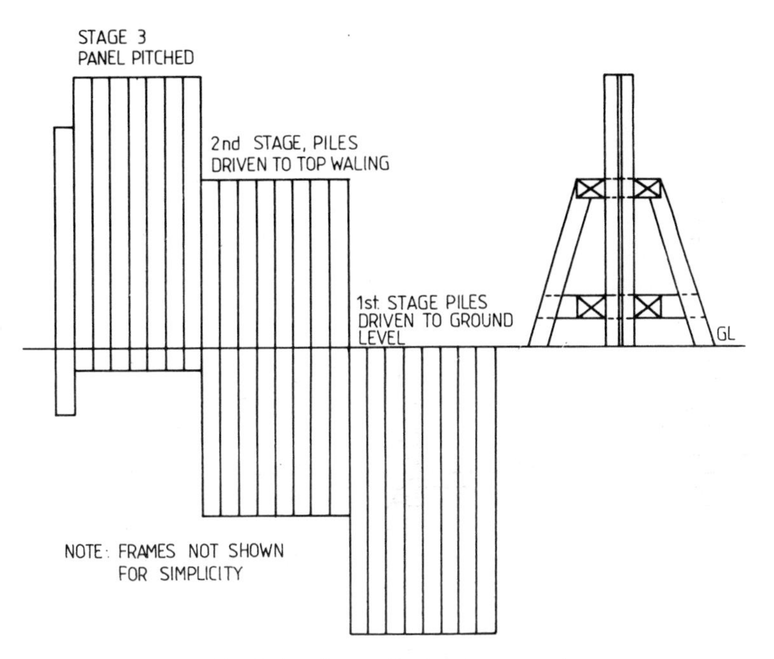

Fig. 8.40 Sheet pile driving procedure

Piling problems

(i) There is a tendency for the play in the interlock to cause leaning.

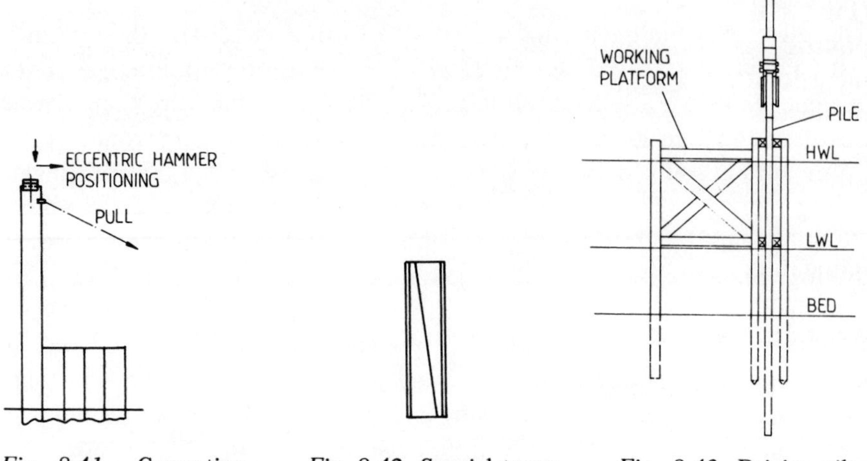

Fig. 8.41 Correcting off-line piles

Fig. 8.42 Special taper sheet piles

Fig. 8.43 Driving piles over water

This can be countered by placing the hammer slightly off centre or alternatively by pulling at the top of the pile (Fig. 8.41). If the leaning cannot be eliminated then the only recourse is to obtain special taper piles (Fig. 8.42) which may involve considerable procurement delays.

(ii) Piles also tend to lean inwards during driving, and this can be countered by providing spacer blocks (Fig. 8.43) between the waling and pile face.

(iii) When driving piles in soft material such as clay, previously driven adjacent piles might be dragged down. This can be prevented by cross-bolting the piles or bolting them to the waling. Overdriven piles should be jacked back into position.

(iv) Limited head room may be overcome by driving short length piles and welding on extension pieces.

(v) When driving piles below the water level, hammers can be equipped to work underwater.

(vi) Piles often get damaged during driving, and also, because it is seldom possible to drive all the piles in a row to exactly the same level, sufficient pile should be left to allow burning off at 100 mm below the damaged parts.

Piling over water

If a temporary platform can be erected to support a crane, then piling may be carried out using a rope-suspended hammer with the platform substructure acting as the guide frame for the panels of sheet piles (Fig. 8.43). Alternatively, where manoeuvrability is required, a pontoon-mounted pile frame or crane fitted with hanging leaders, incorporating a long dolly between the pile cap and hammer, may be preferred. Positioning must then be obtained by winching from temporary anchorages. Currents and tide unfortunately pose serious problems for this latter method.

Piling data

Weight of hammer

W = weight of hammer
P = weight of pile

Concrete piles	$W \simeq P$
Timber piles	$W \simeq 2P$
Steel tubes and Sheet piles	$W = 0.5P$ to $2.0P$ in dry sand $= 2.0P$ to $2.5P$ in saturated sand $= 2.5P$ to $3.0P$ in clay
I-shaped piles	$W = P$

Hammer/pile type selection

Timber piles – drop or single-acting hammer
Concrete piles – drop or single-acting hammer with fall or stroke less than 0.5 m
Steel or sheet piles – double-acting air hammers. For silent driving, vibration or hydraulic methods may be adopted depending upon the type of soil.

Note: These are the preferred types of hammer, but the demand for a very heavy hammer or for a diesel hammer giving high impact energy may be overriding factors with heavy piles and/or hard soils.

Preferred hammer/soil type selection

Light soils	Loose sands and gravels	– Double-acting air hammer
	Soft clay →	Drop or single-acting air hammer
Medium soils	Medium dense sands, Fine gravels	– Double-acting air hammer
	Stiff clay →	Drop or single-acting air hammer

Hard soils	Dense sand →	Double-acting air hammer
	Coarse gravel →	Single-acting air hammer or diesel hammer
	Very stiff clay →	Drop or single-acting air hammer

Note: For heavy piles, the range of double-acting air hammers may be too small and single-acting hammers may be the only alternatives.

Production rates of pile-driving

Many factors must be included, such as the time needed for setting up the pile frame or rigs, the skill of the work gang, weather losses, soil types, length of pile, type of pile hammer, etc., and it is therefore extremely hazardous to recommend typical driving rates. However, as a very general guide the following data may help in planning piling operations.

		Single piles (m/h)	*Sheet piles* (m^2/h)
Light soils	using the appropriate hammer	5 - 15	3 - 10
Medium soils	using the appropriate hammer	3 - 9	2 - 6
Heavy soils	using the appropriate hammer	1.5 - 5	1 - 3

For sheet piling using the Taywood Pilemaster, up to 20 m^2/h can be achieved in soft ground. Using vibratory methods, over 30 m^2/h have often been recorded. Extracting piles with a compressed-air pile extractor requires about half the time required for driving.

Piling gang

One ganger, one crane driver or winch operator, and two labourers to grease, prepare and handle piles, fix support brackets, etc. An additional labourer may be needed to assist with assembly when using a temporary timber guide frame. Transport should also be provided to bring the piles from the storage compound.

References

British Steel Corporation (1970) Larssen Sheet Piling, BSC Publications.
British Steel Corporation Piling Handbook (1983) BSC Publications.
British Steel Piling Company (1983) B.S.P. Pocket book. Ipswich.
British Steel Piling Company (1980) BSP-Muller vibrators.
British Steel Piling Company (1975) Frodingham Steel Piling.
BSP International Ltd (1983) Frodingham straight web piling.

BSP International Ltd (1983) Pile driving equipment and piling systems.
BSP International Ltd (1983) Steel trench sheeting.
Burland, J.B. and Potts, D.M. (1981) 'The overall stability of free and embedded cantilever walls.' *Ground Engineering*. July.
CP 2004 (1970) – Foundations. British Standards Institute.
Dawson Ltd. (1980) Dawson Quiet Piling, Dawson Ground Anchors, Dawson Ground Release Shackles, Dawson Sheet Pile Threader, Luton.
Delmag Company, (1983) Pile driving equipment and piling systems, Esslingen, Germany.
Foster Ltd. Foster Vibro driver/extractor, 1979.
Krupp Company (1975) – Vertical Sheeting, Sheet piling booklets. Düsseldorf, Germany.
Millbank, D. (1973) 'Steel sheet piling today.' *Civil Engineering and Public Works Review.*
Noise and vibration from piling operations (1980) CIRIA Report PG9.
Page, E.W.M. and Semple, W. Silent and vibration-free pile driving. Proceedings of the Institution of Civil Engineers, 1968, **41** (Nov.) and 1969, **43** (June).
PTC Company (1982) Vibroforceurs, Paris.
Taylor-Woodrow Ltd. (1979) The Taywood Pilemaster, Southall, Middx.
Tomlinson, M.J. (1963) *Foundation design and construction.* Pitman, London.
Vulcan Iron Works Inc. (1983) Onshore and Offshore Pile Hammer, Chattanooga, Tennessee, USA.

CHAPTER NINE

Shoring systems

Introduction

When carrying out excavation work on a congested construction site the sides of an excavation frequently require shoring (Fig. 9.1). When space is not at a premium, however, the sides are often left open and simply battered back (Fig. 9.2).

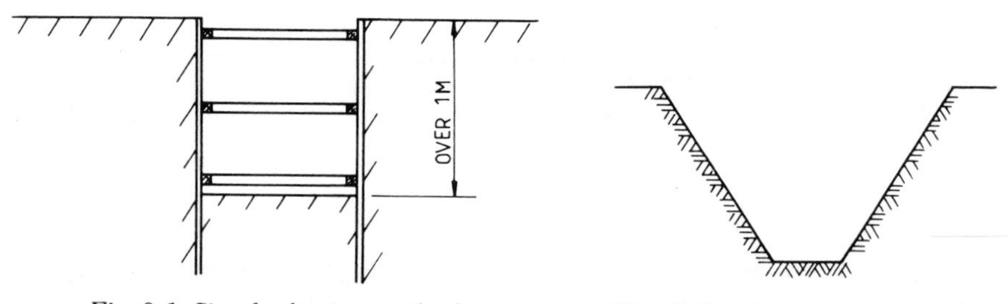

Fig. 9.1 Simple shoring method

Fig. 9.2 Open cut excavation

The choice of method depends upon technical factors, such as the depth of the foundation, the available space, the nature and permeability of the soil, the depth of the water table and economic considerations related to the period the excavation must be left open, the availability of labour, plant and materials and the construction programme. Table 9.1 provides an indication of the methods required in various ground conditions.

Open excavations (*Fig. 9.3*)

Slopes excavated in *dry sands* and *gravels* quickly find their natural angle of repose. The angle largely depends upon the degree of compaction, and shape of the grains.

Moist sands tend to give a misleading picture as they will often stand almost vertically for a short time, but eventually the face either breaks

Table 9.1 GUIDE TO GROUND CLASSIFICATION AND SUGGESTED SHORING METHOD (BASED ON AN ORIGINAL BY THE CITB)

TRENCH DEPTH	UP TO 5FT	5FT –15FT	OVER 15FT	SOUND ROCK	FISSURED ROCK	FIRM AND STIFF CLAYS	GRAVELS SANDS: COMPACT	GRAVELS SANDS: SLIGHTLY CEMENTED	FIRM PEAT	GRAVELS SANDS: BELOW WATER TABLE	GRAVELS SANDS: LOOSE	SOFT CLAYS AND SILTS	SOFT PEAT
SHALLOW				A	A	B-C	A	A	A	C	C	C	C
MEDIUM				A	A	B-C	B	B	C	C	C	C	C
DEEP				A	B	C	C	C	C	C	C	C	C

A. NO SUPPORT MAY BE NECESSARY. B. OPEN SHEETING.
C. CLOSE SHEETING OR BETTER

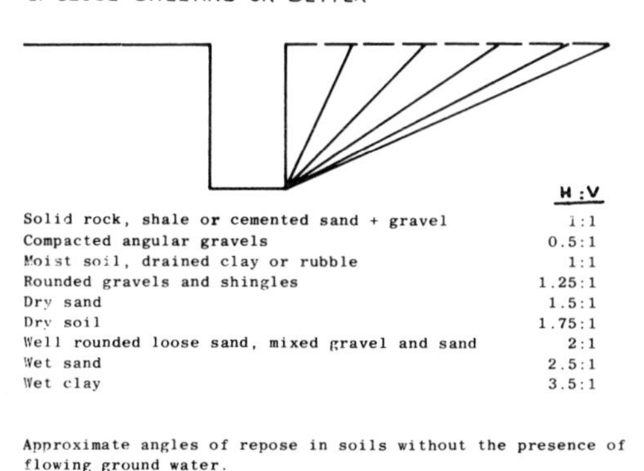

Fig. 9.3 Suggested batters for open cut excavations

up in lumps, or slumps to take up a natural resting position. Obviously, such soils are particularly dangerous, for example in pipe laying work, where unsuspecting workmen have frequently been trapped in collapsed trenches.

Water-bearing sands pose different problems. Here, the water flows into the excavation causing erosion of the toe of the slope, resulting in

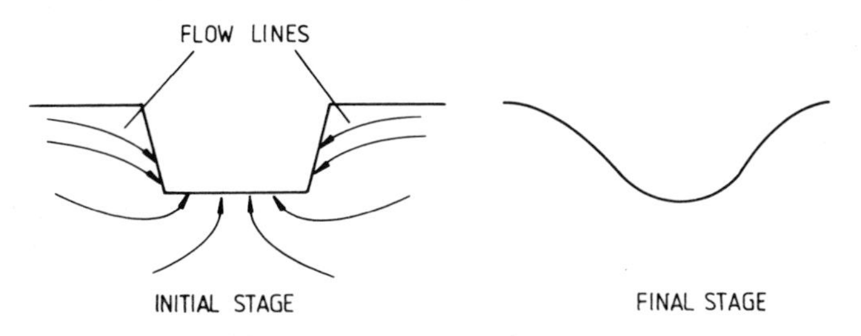

Fig. 9.4 Effect of ground water in open cut excavations

progressive collapse as shown in Fig. 9.4.

Silty sands, when dry, will stand almost vertically, expecially if the silt is slightly cohesive. However, wet silts are extremely troublesome and suffer similar problems to water-bearing sands. Excavation in this material is extremely tedious as the soil flows into the excavation almost as quickly as it is removed, thus producing a very flat shape. Generally, open excavations in such material are impracticable.

Cohesive soils such as clay soil can theoretically be shown to stand vertically at a height dependent upon the cohesive strength. However, fissured clays, after excavation, tend to slide along these fissure planes. Even when fissures are not present, drying out on the surface causes shrinkage cracking, which later fills with water. Hydrostatic pressure then causes breaking away of sections of the clay face (Fig. 9.5).

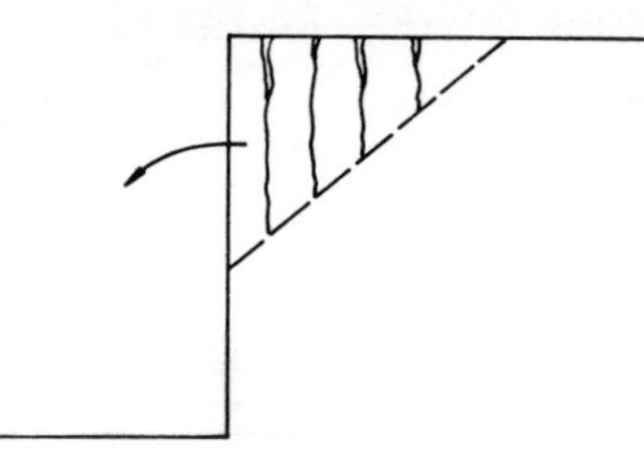

Fig. 9.5 Fissuring of clay soils

In most clays the safest action is to slope back the sides. The required section can be determined from slip circle calculations (Fig. 9.6).

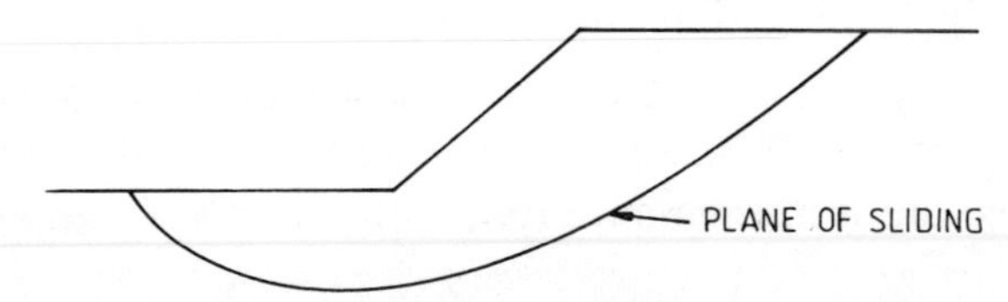

Fig. 9.6 Typical plane of sliding of a failed embankment in a clay soil.

Closed excavations

Trenches

Self support (*Fig. 9.2*)

Safety regulations do not require a trench to be lined for excavations shallower than 1.2 m; however, for greater depths a shoring method is recommended unless the sides are battened back.

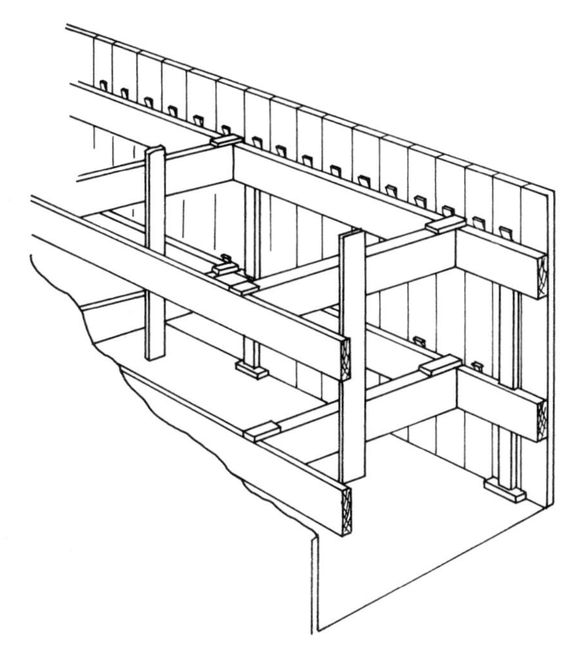

Fig. 9.7 Planking and strutting for timbered trenches

Planks and struts

Planking and strutting was originally the accepted method of trench shoring. Figure 9.7 illustrates close boarding for applications in running soil and generally poor ground. Less planking may be possible in more stable conditions, for example when the boards are at about 2 m spacings the method is usually referred to as open timbering. Today, because of the high cost of timber and specialist skills required, other methods such as trench sheeting, lining panels and box systems are gaining popularity.

Trench sheets and props (*Fig. 9.8*)

The use of trench sheets is a very economical method of support. The installation procedure (Fig. 9.9) is similar to that for driving a panel of sheet piles.

The trench is first excavated to a depth of 0.5 – 1 m to steady the piles. A panel of sheets is pitched along opposite sides of the trench and propped apart. Pile-driving by means of a light double-acting hammer then proceeds to just below the next planned working level, excavation follows to this depth and further propping is installed. The process is repeated until the full depth of trench is obtained.

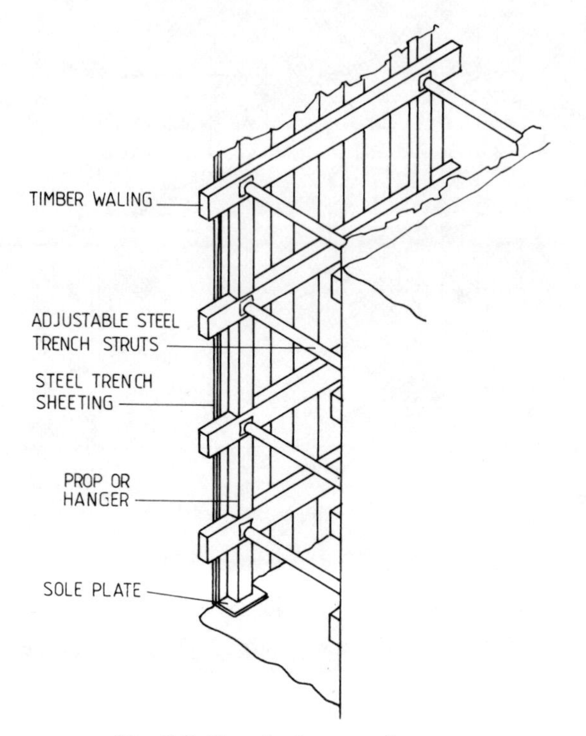

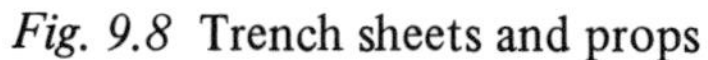

Fig. 9.8 Trench sheets and props

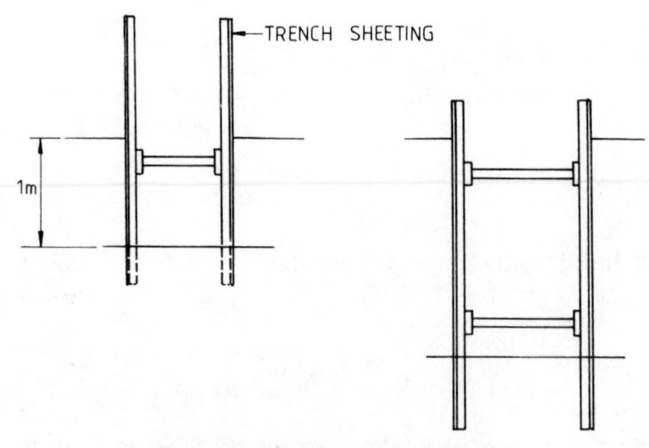

Fig. 9.9 Installing trench sheets

Trench sheets and hydraulic vertical shores (*Fig. 9.10*)

The system is primarily an alternative to open timbering and comprises integral hydraulically-operated props and support plates (Fig. 9.11).

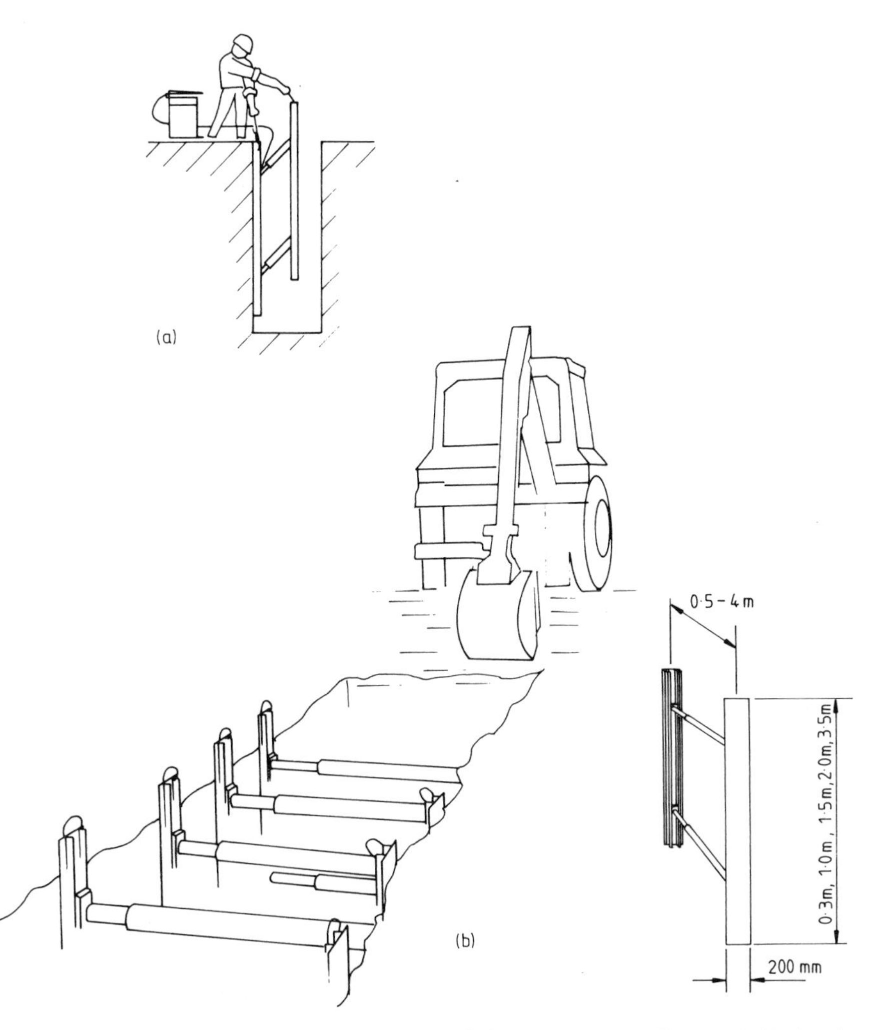

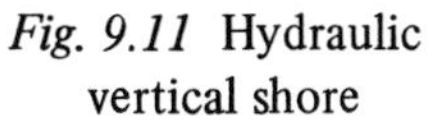

Fig. 9.10 Installing hydraulic vertical shores

Fig. 9.11 Hydraulic vertical shore

Care must be taken during installation to ensure that the cylinders are perpendicular to the trench sides, otherwise on application of hydraulic pressure, one side of the frame may simply slide upwards. The hydraulic fluid is pumped in through flexible tubing from a hand pump operated at surface level. The whole system is very versatile and varying trench widths can be accommodated. The equipment is made in aluminium and is therefore light to handle.

In hard stable ground, such as soft rock, the frames may be spaced up to 2 m apart. In stiff clay and other stable cohesive soils, 1 m spacings may be sufficient. However, in sands and gravels 200 – 300 mm or less is often required. In all situations the ground should be free from excessive running water to avoid instability problems.

Trench sheets and hydraulic horizontal shores (walers) (*Fig. 9.12*)

Horizontal shores are designed for use when close sheeting is required. Removing the sheets and backfilling is made much easier compared to a conventional propping method by the control over releasing the

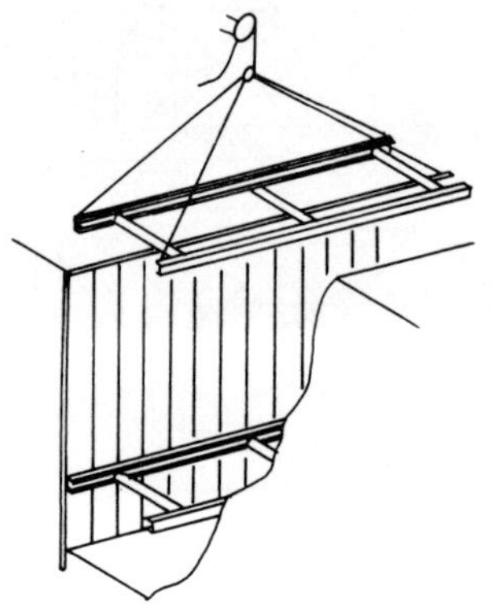

(a)

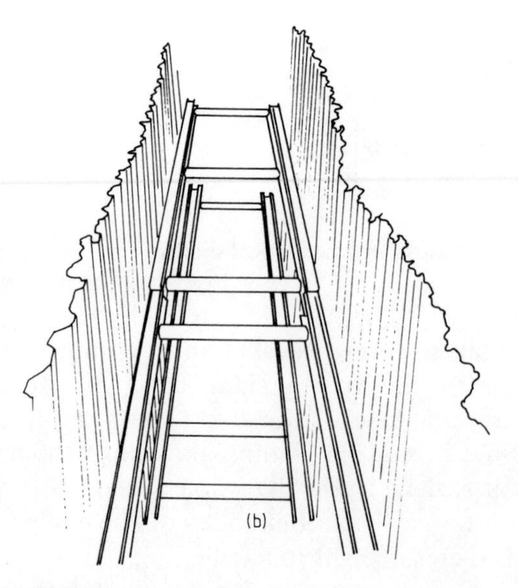

(b)

Fig. 9.12 Installing hydraulic horizontal shores (walers)

pressure in the cylinders. This allows the walers to be withdrawn from the trench in stages, thereby enabling backfilling to take place in safety.

The walers are fabricated in one piece aluminium units, which can be positioned from above (Fig. 9.13).

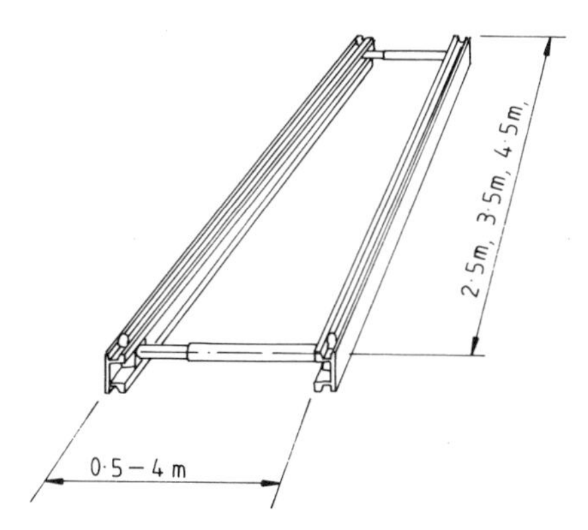

Fig. 9.13 Hydraulic horizontal waler

Lengths are available from 2 - 5 m, and units may be coupled together to line a full length of trench.

Drag boxes (*Fig. 9.14*)

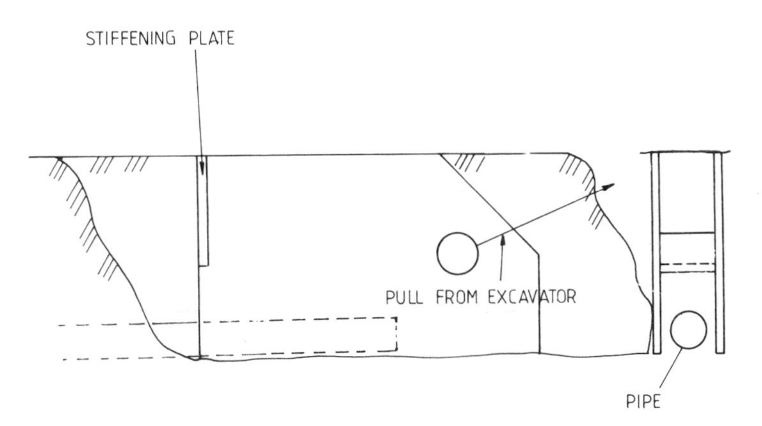

Fig. 9.14 Drag box

Drag boxes are used as safety boxes for operatives rather than trench supports. The method of operation requires the trench to be cut slightly wider than the box, which is then pulled forward into position by the excavator. This method is unsuitable when ground movement must be avoided, or where services cross the line of the trench. It is both a safe and quick procedure but requires a large excavator to handle the box.

Trench boxes (*Fig. 9.15*)

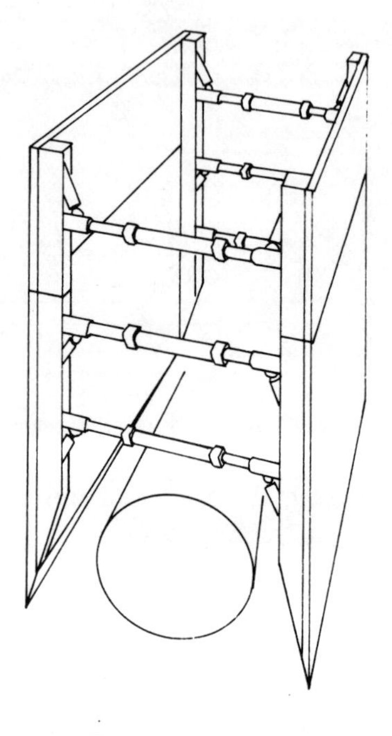

Fig. 9.15 Trench box

For genuine trench shoring, trench boxes have recently appeared as an alternative to sheeting. The trench box is a modular system composed of two support walls separated by props. A method of installation is demonstrated in Fig. 9.16. Contractors have generally found that three boxes are sufficient to operate an efficient cycle of work - one box going down with the excavating, a second box already founded to provide protection for pipe installation and the third box coming up as backfilling proceeds.

A large excavator capable of lifting is necessary to handle the boxes and achieve economic rates of digging and installation.

Slide rails

The Krings Company has developed the trench box system a stage further to incorporate slide posts driven ahead of the side plates to act

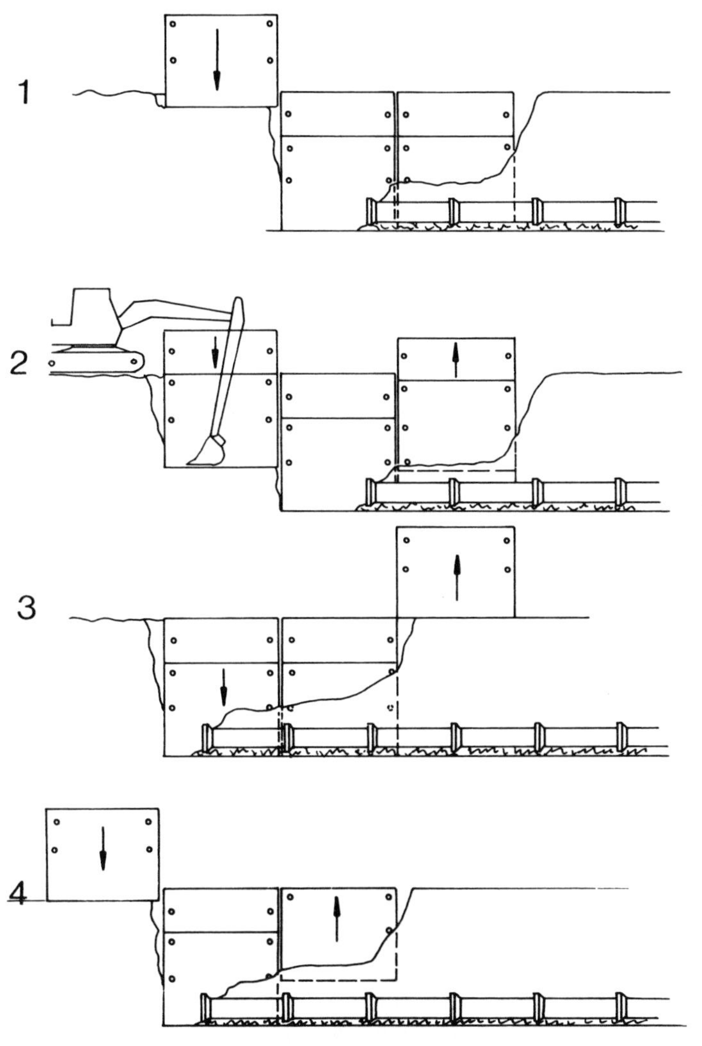

Fig. 9.16 Installing trench boxes

as guides (Fig. 9.17). The manufacturers claim that this method overcomes the problems of withdrawal in granular soils.

Other specialist systems are available, for example the Supershore by Shorco and the Ramshore by Hudswell Morrice, but have been little used in the U.K.

General comments on trench support

Trench sheets are relatively thin in section and therefore leave only a small void on extraction. However, bowing and buckling may take place

Table 9.2 Comparison of trench shoring systems

System	Comparison
Slide rails	*Advantages* (i) Operatives not required to enter unsupported trench (ii) Provides a comparatively safe, clean and dry working area for pipe laying (iii) Excavation and shoring in one operation (iv) Requires a lower rated excavator than boxes *Disadvantages* (i) Capital outlay (ii) Cross services are difficult to deal with (iii) Installation becomes difficult if the system gets out of line
Trench boxes	*Advantages* (i) Operatives not required to enter unsupported trench (ii) Provides a safe, clean and dry working area for pipe laying (iii) Increased output (iv) Reduction in gang size *Disadvantages* (i) Large capital outlay (ii) Boxes are heavy and require increased capacity of excavator
Trench sheets	*Advantages* (i) The method is well known by planners and workmen alike – for many contractors it remains the 'only' solution to trench support (ii) Trench sheets and props are easy to transport, require little storage space, and are available from numerous plant hire companies up and down the country (iii) The method is versatile, varying trench widths and depths are easily accommodated (iv) Differing ground conditions are catered for by varying the spacing of the sheets (v) They are relatively cheap either to buy or hire (vi) Cross services create few problems *Disadvantages* (i) It takes a considerable amount of time and labour to install and extract (ii) The lowering of pipes through a mass of trench props is difficult (iii) Workmen have to enter unsupported trenches to install and remove the supports

	(iii) Cross services are difficult to deal with (iv) There is a gap between adjacent boxes through which water and soil can flow; this is usually bridged by using a trench sheet
Drag boxes	*Advantages* (i) Very high output rates (ii) Operatives do not have to enter unsupported trench *Disadvantages* (i) Large excavator needed for pulling the drag box along, and a crane required for handling (ii) Drag boxes have fixed dimensions (iii) Large capital outlay (iv) Cross services are a major problem
Hydraulic shores	*Advantages* (i) Installation from above ground (ii) Quick to install and remove (iii) Fewer labourers needed in gang (iv) Cross services are not a problem *Disadvantages* (i) More expensive than sheets and props (ii) The systems require maintenance

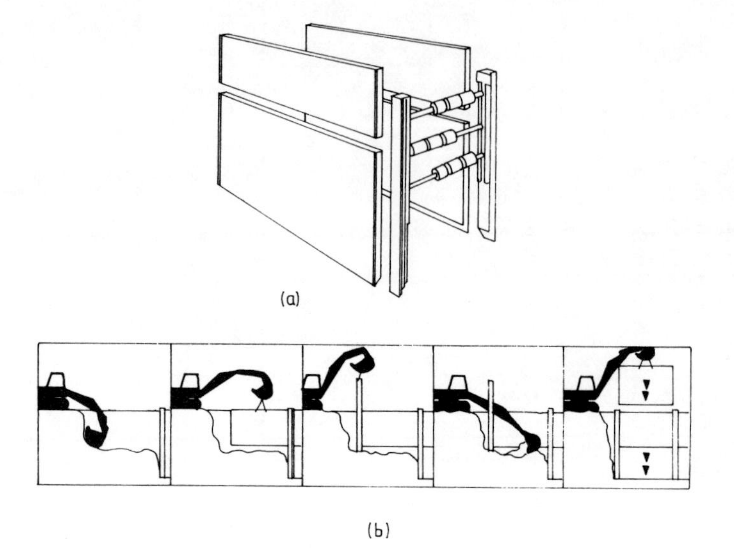

Fig. 9.17 Slide rails

and some ground movement is then unavoidable. Drag boxes provide adequate protection for the operative, but excessive ground movement is likely. Trench boxes and slide rails avoid most of the movement problems but extraction and ground services can cause severe difficulties.

The presence of groundwater may cause problems for all the methods, particularly in running sands and silts subject to piping (see page 108). Production rates in such conditions might be only a fraction of normal, because material may continue to flow into the trench while excavation progresses (Fig. 9.18). In such circumstances a de-watering method should be installed.

A summary of the merits of the various systems is given in Table 9.2.

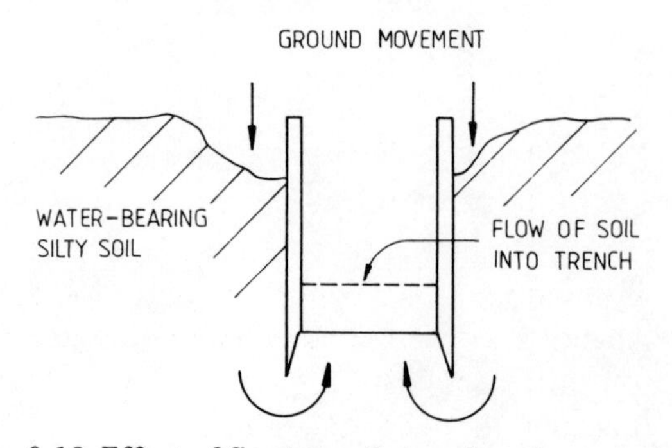

Fig. 9.18 Effect of flowing soil on a shored excavation

Table 9.3 Output rate of various trench support systems in m^2/h of lined trench

System	Soil type: Strong		Medium		Loose	
Depth (m)	2 - 4	4 - 6	2 - 4	4 - 6	2 - 4	4 - 6
Traditional trench sheeting	10	14	6	8	2.5	3.5
Hydraulic walers and trench sheets	*	*	*	10	*	*
Trench box	*	17	10	15	*	4.0
Slide rails	*	6	*	*	2.5	*
Drag box	*	*	8	*	3.5	*

Note: * No data was available

Output data

Short has collected output data (Table 9.3) from a number of contracting firms. The units are expressed in m^2/h of trench sides supported, rather than m^3/h of trench excavated, because installation of the support is usually the slower activity.

Recommended plant and labour selection (including pipe layers)

The choice of plant and gang size varies from contractor to contractor for any particular operation. In the example given below, two machines are assumed - one for excavating and the other for backfilling and lifting duties. At present, there is no standard available defining safe working loads for excavators and therefore it is only possible to give approximate machine sizes to provide adequate safe lifting capacity.

There will of course be occasions when the recommendations given will not be valid. For example, when excavating in rock or for particularly shallow trenches. However, they will serve as a guide to new users of the systems.

Trench sheets

One backhoe (with 0.50 m^3 bucket) for excavation

One backhoe (with 0.25 m^3 bucket) for backfilling, placing pipes and pea gravel, and extracting trench sheets. It is common practice on many sites for the backhoe to spend 50% of its time on these tasks and the remainder on different jobs around the site. This is possible because of the machine's good manoeuvrability.

One ganger

Five labourers

Note: Sheet pile-driving equipment may be required in deep trenches (>3 m).

Trench boxes

There are two possible selections:

(i) One backhoe 0.5 m^3 capacity for excavation
One backhoe 0.5 m^3 capacity (15 - 18 tonnes class) for backfilling, etc. plus extracting and placing the boxes. This machine is slightly underrated for the task and, therefore, the boxes have to be removed in individual pieces (i.e. the top sections first and then the base unit). In granular material it has been known for a machine of this capacity to be unable to extract boxes because of the high ground pressures.

(ii) One backhoe 0.5 m^3 capacity
One backhoe (with 2.5 m^3 bucket, 35 - 40 tonnes class) for backfilling etc. plus extracting and placing the boxes.
3 sets of trench boxes. Manufacturers recommend the use of four sets, but contractors find they can cope adequately with three.
One ganger
Four labourers

The second plant selection is recommended, although it is more expensive, since it does give higher output and reduces the possibility of the boxes getting stuck.

Slide Rails

One backhoe 0.5 m^3 capacity for excavation
One backhoe 0.5 m^3 capacity for backfilling etc., plus lifting and placing the slide rails
Three sets of linings
One ganger
Four labourers

Hydraulic Shores

One backhoe 0.5 m^3 capacity for excavation
One backhoe 0.25 m^3 capacity (for 50% of the time)
One ganger
Four labourers

In strong ground conditions vertical aluminium shores are recommended, and in mixed or granular soils, trench sheets with horizontal aluminium waler frames. Contractors find that 3 m long frames provide enough working length. The number of layers of frames depends upon the depth of trench and the soil pressure.

Note: (i) With all systems, compaction plant and pumping equipment may be required.
(ii) For trench boxes and linings, systems are at present available to provide supported excavations up to approximately 6 m deep and 5 m wide. This restriction is not applicable to sheet piling but careful design is required because of the considerable ground forces involved.

Coffer-dams

Sheet piled coffer-dam

A coffer-dam is a 'watertight' structure which allows foundations to be constructed in the 'dry', for example bridge piers. Nowadays however the term also covers land-based operations and the coffer-dam is a common feature in foundation works. Sheet piling is a popular material used in the construction of medium to large coffer-dams, while trench sheeting or even timber boarding is more appropriate for shallow excavations (Fig. 9.19). A serious hindrance with these methods is the

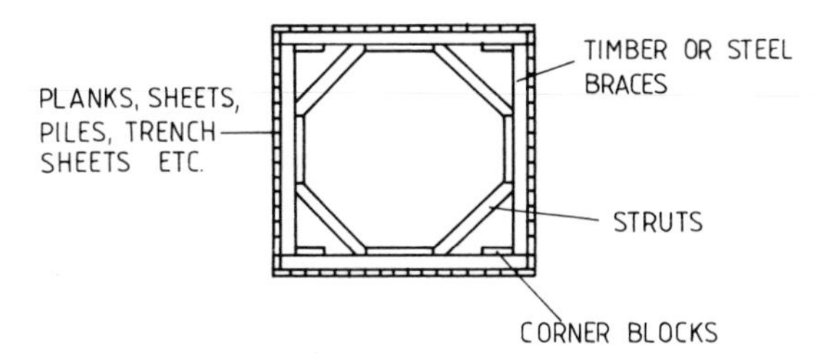

Fig. 9.19 Timbered coffer-dam

fairly cumbersome bracing required to hold the piles in position. However, modern developments in hydraulically-pressurised modular aluminium frames are proving quite successful where the sides do not exceed about 5 m, (Fig. 9.20). For more general applications involving irregular shapes etc. however, walings are normally fabricated from RSJs or similar. Indeed, individual sheet piles may also be strengthened in a composite construction.

An alternative and sometimes cheaper method is to prop from the base of the excavation as shown in Fig. 9.21 but obviously this can only be progressively carried out as the excavation is deepened, if excessive cantilevering is to be avoided.

Another alternative is to build a circular coffer-dam and form the

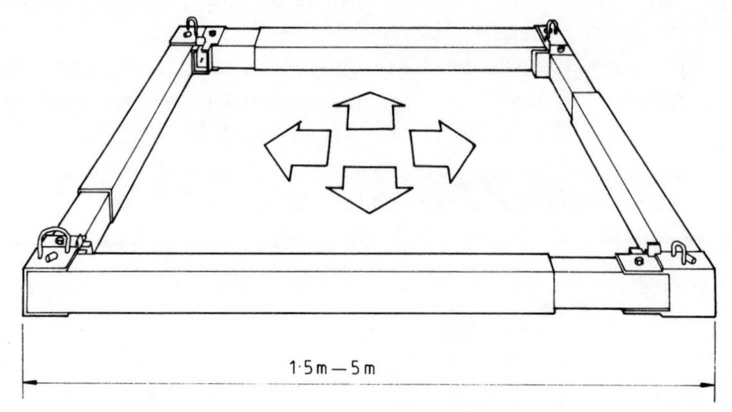

(a)

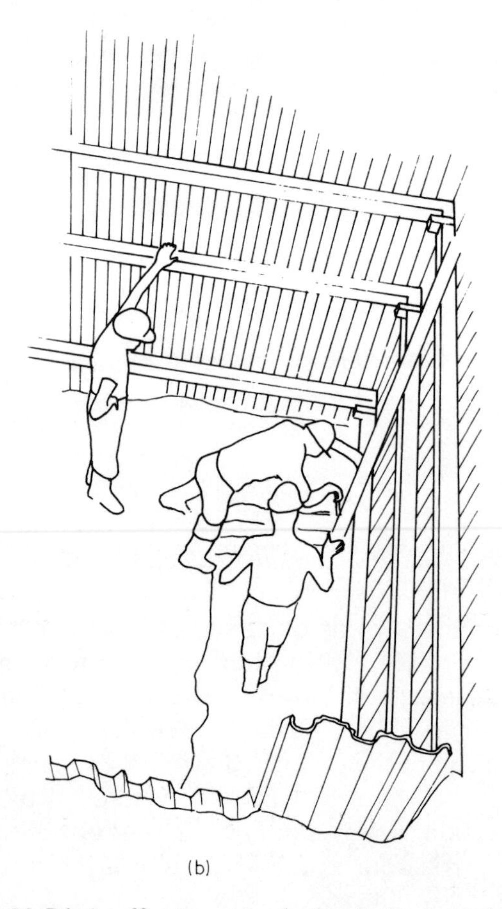

(b)

Fig. 9.20 Piled coffer-dam using hydraulic horizontal walers

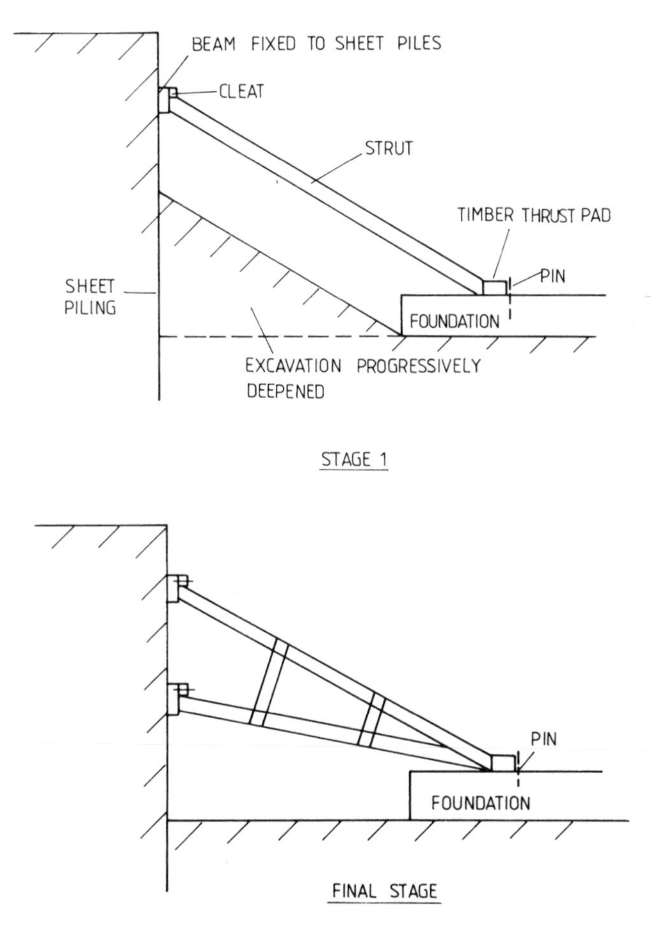

Fig. 9.21 Propped piled wall

walings in reinforced concrete, thereby utilising the advantage of ring compression as shown in Fig. 9.22.

Methods of sheet pile installation are discussed in Chapter 8.

Design of sheet piled walls

The design of sheet pile walls and coffer-dams depends upon many factors, including the depth of excavation, water table, soil type, surcharge, propping the tying arrangement etc. To demonstrate all these variables is beyond the scope of this book and the reader is directed to the BSP Pocketbook.

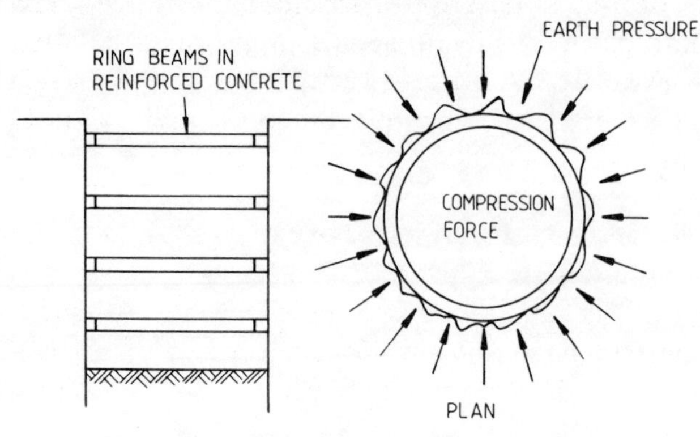

Fig. 9.22 Coffer-damming using ring beams

Reinforced concrete piled walling

In an attempt to reduce the cost of shoring, support systems have been designed as part of the permanent works. For example, the first methods involved the use of in situ concrete piles cast to form a continuous wall. Furthermore, the row of piles could be given a facing of reinforced concrete to improve the surface appearance (Fig. 9.23).

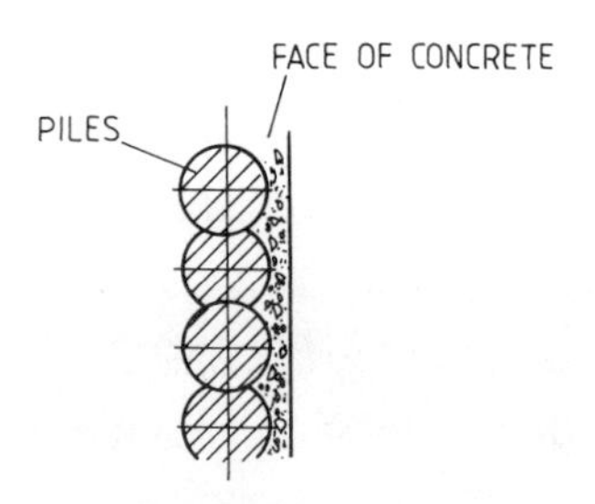

Fig. 9.23 Forming a finished surface with reinforced concrete piled wall

Types of wall

(i) The simplest and cheapest method is the single row of tangentially touching piles (Fig. 9.24) (contiguous piles). The pile is constructed either by boring or by using a grabbing bucket. In poor or water-bearing ground, lining or bentonite slurry may be necessary to support

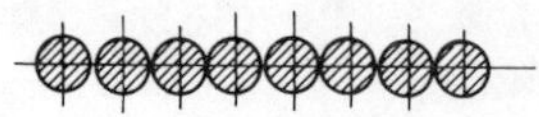

Fig. 9.24 Simple reinforced concrete piled walling

the walls of the hole. Reinforcement is positioned and concrete tremied into place. Finally, the ground in front of the completed row of piles is excavated. A watertight connection between adjacent piles cannot be achieved, as in practice the positioning of each pile often leaves a 50 - 100 mm gap. A de-watering system may therefore be necessary.

(ii) The above method can be improved to produce a closed joint by constructing alternate piles, followed by the intermediate ones as illustrated in Fig. 9.25. This is achieved by boring or cutting away part of the first phase piles while the concrete is still weak (i.e. less than 5 N/mm^2). The system is usually referred to as *Secant-pile* walling.

Fig. 9.25 Secant concrete piled walling

Continuous diaphragm walls (icos walls)

This system developed from the technique of forming a coffer-dam from a row of bored and cast in situ piles sunk in close contact with one another.

Method of construction (*Fig. 9.26*)

Step 1: Construction guide walls

A trench is excavated about 1 m deep along the line of the wall, wide enough to accommodate the wall and to give enough clearance for the grab. The inner face only is shuttered.

Step 2: Excavate first and subsequent panels

(i) A slot is excavated to the requisite depth between the guide walls to form the first panel.
(ii) Slurry (3 - 10% powdered bentonite in water) is then pumped into the trench.

Step 3: Place cage of reinforcement

(i) The trench is cleared of slurry/soil sediment at the base.
(ii) The ends of any abutting panel are scraped clean by the teeth of the excavator bucket.
(iii) Steel stop ends are inserted.
(iv) A cage of reinforcement fitted with spacer blocks is assembled.
(v) Reinforcement is lowered into the slurry.

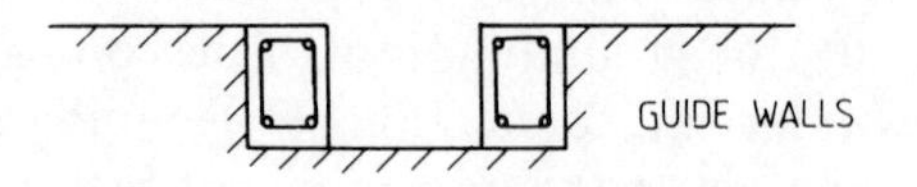

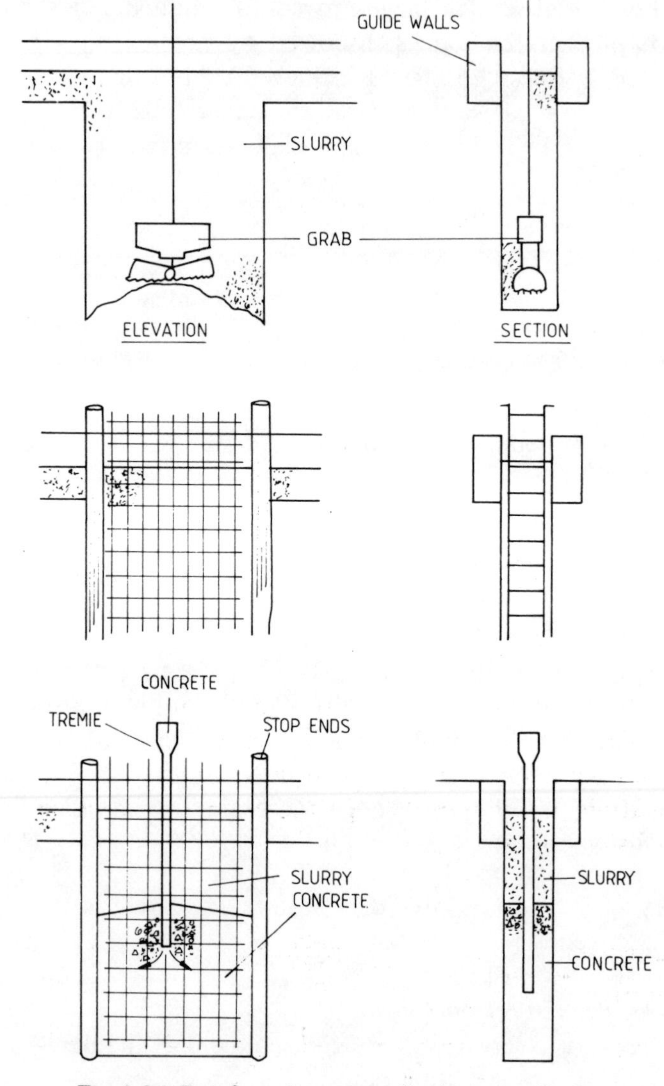

Fig. 9.26 Reinforced concrete diaphragm walling

Step 4: Tremie concreting

High slump concrete is finally placed in the trench using a tremie tube. The complete sequence of operations for subsequent panels is shown in Fig. 9.27.

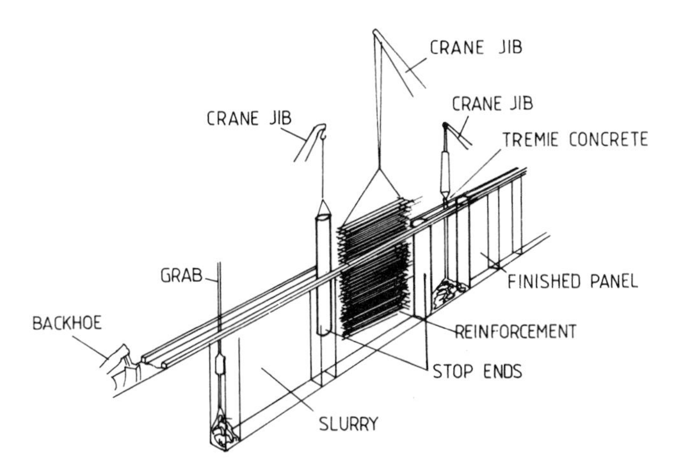

Fig. 9.27 Installing procedure for diaphragm walling

Construction guide-lines

Construct panels in	2 - 6 m lengths
Wall thickness	0.5 - 1 m
Trench depth	Up to 35 - 40 m
Vertical accuracy	1 in 80

Mixing and placing the slurry

The slurry is made from bentonite clay delivered to site as a dry powder and then mixed with water.

A typical systematic mixing set-up is shown in Fig. 9.28. The equipment consists of a mixing hopper with paddle wheels rotating at about 250 r.p.m. The mix is passed through a centrifugal pump rotating at 1000 - 1500 r.p.m. at the base of the drum and returned. The result is a rapid circulation of the mix through the hopper tank which produces the colloidalising effect.

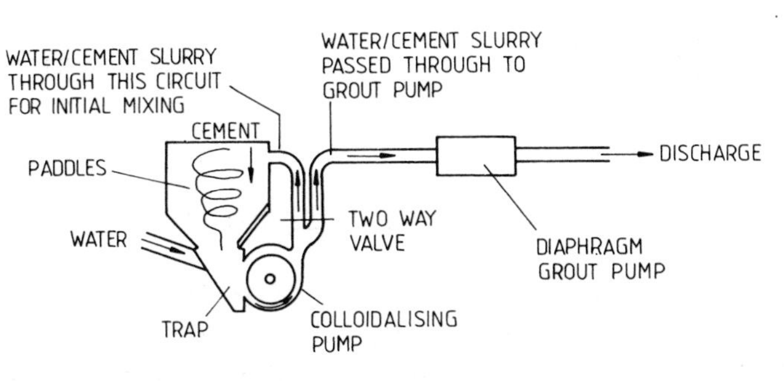

Fig. 9.28 Grout or bentonite mixer

The mixture is finally pumped to position with a piston pump. Because of the thixotropic properties of bentonite it may be necessary to place an agitating hopper between the mixer and delivery pumps when the material is not required continuously. Typical production data are given in Table 9.4.

Table 9.4 Data for low pressure grouting equipment

Output (*l*/m)	50	75	100
Hopper capacity (*l*)	10	10	10
Pump discharge pressure* (bar)	7	7	7
Power required (kW)	3	3	5

**Note:* For high pressure grouting a piston pump capable of producing 70 bar may be required (see Chapter 7).

Bentonite properties

The solids in the bentonite fill the pores of the soil to form a weak membrane. Because the density of the slurry is greater than the particle/water soil density, outward hydrostatic pressure forces the membrane against the sides of the excavation. Thus, support is provided and the hydrostatic head in the water table balanced.

The density of the fluid may be increased by raising the bentonite concentration as follows:

4% by weight bentonite – 1022 kg/m^3 density is suitable in stiff clay

10% by weight bentonite – 1060 kg/m^3 density is suitable in coarse soils.

Re-cycling of bentonite slurry

The bentonite slurry gets contaminated by soil particles during excavation which, if excessive, destroy its properties. However, to reduce costs, for example in a long section of walling, the displaced bentonite is led into a tank and allowed to settle. The top layer is then returned to the trench. Waste slurry is usually transported to a tip in tanker trucks for disposal.

Testing the slurry

(i) *Viscosity*

A rough viscosity measurement may be taken on site using the Marsh Funnel (Fig. 9.29).

The time of outflow of 0.95 ℓ of slurry is compared to that of water.

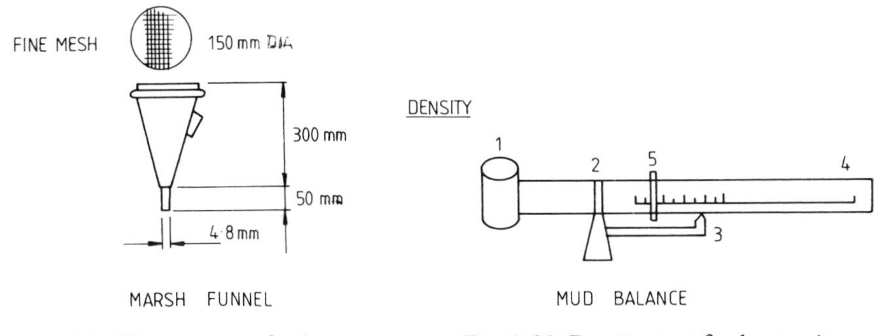

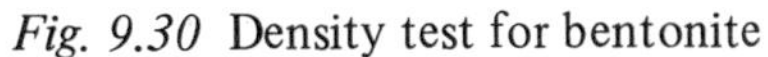

Fig. 9.29 Viscosity test for bentonite

Fig. 9.30 Density test for bentonite

(ii) *Density*
A given volume of slurry is weighted on a balance beam (Fig. 9.30). For a given bentonite concentration the difference between the density of fresh slurry and that measured on the balance will be due to the contamination by soil particles.
Note: Experience is at present the best guide to the most suitable bentonite concentration and the point at which slurry should be treated for re-use or disposed of.

Applications

Bentonite diaphragm walling is particularly suitable when combinations of both temporary and permanent support are included in the design of the structure, for example

(i) basement construction (Fig. 9.31)
(ii) pumping chambers } (Fig. 9.32)
(iii) underground tanks } (Fig. 9.32)
(iv) coastal defence works and river walls (Fig. 9.33)
(v) retaining walls (Fig. 9.34)

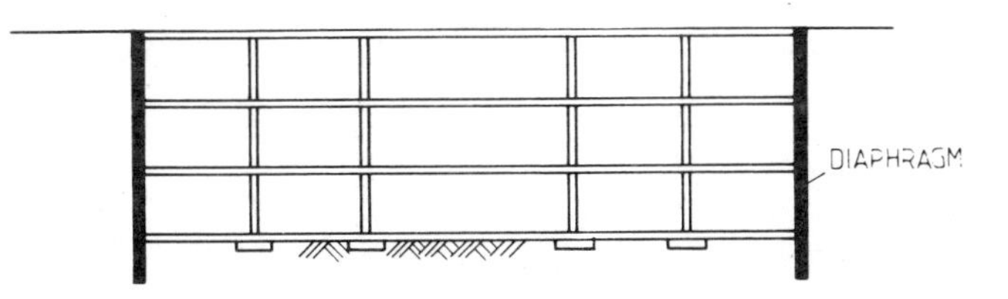

Fig. 9.31 Diaphragm walling for a basement building

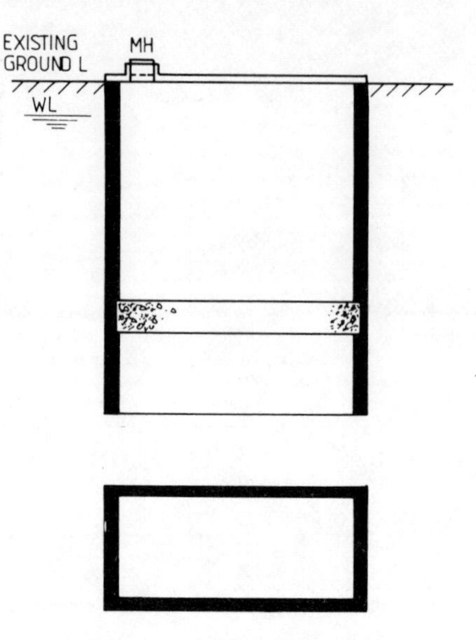

Fig. 9.32 Diaphragm walling for a pumping chamber

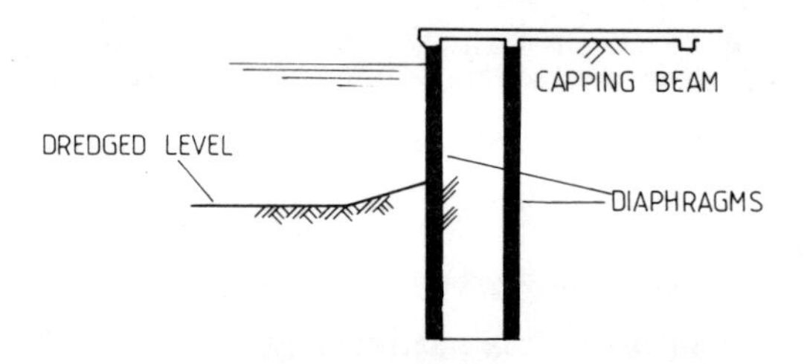

Fig. 9.33 Diaphragm walling for coastal defences

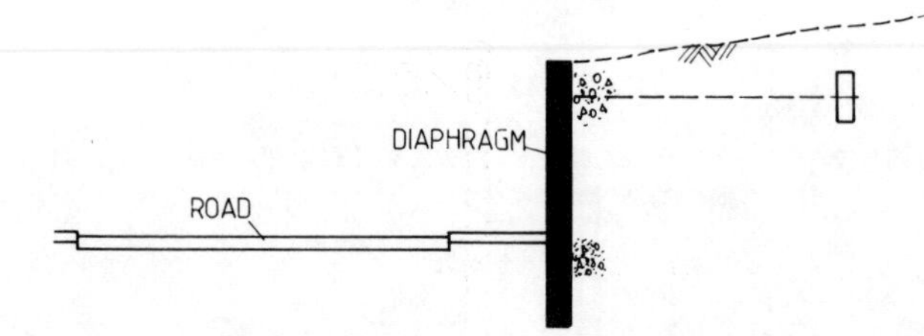

Fig. 9.34 Diaphragm walling as a retaining wall

Injected membranes

In water-bearing ground, a continuous impermeable membrane installed around the outside of a proposed excavation to provide a dry working area (Fig. 9.35) is sometimes a more economic solution than using sheet

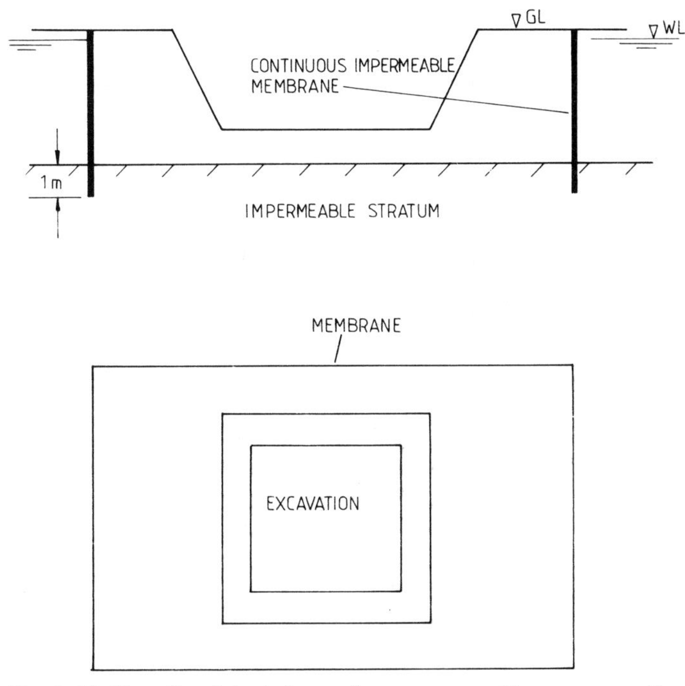

Fig. 9.35 Use of an injected membrane surrounding an excavation

piling, diaphragms, etc. The method, originally developed by Études et Travaux Fondations of Toulouse, involves a series of H section piles. When approx. seven units are in place, the rearmost pile is withdrawn while the void beneath is injected with grout at 2 – 5 bar pressure through a 20 mm diameter pipe running along the face of the pile (Fig. 9.36). Extraction is usually performed with hydraulic rams, as the

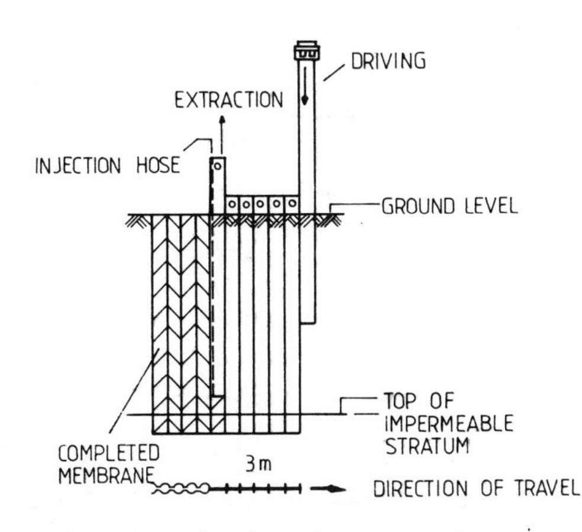

Fig. 9.36 Installing injected membrane

conventional hammer-type extractor method might disturb and damage the grouted membrane. The extracted pile is then transferred, re-driven as the lead pile, and the process continued until the seal is completed.

The grout used is a clay/cement mix, e.g.

45% clay
35% water
20% cement

Approximately 250 – 300 kg of grout is required per m² of membrane. Output – 0.25 hours per m² of membrane may be required to drive and extract the piles, carry out the injection of grout, etc.

Ground freezing

Ground freezing can be used as a means of ground support, water cut-off or combinations of both, as illustrated by examples of a coffer-dam and tunnel heading shown in Fig. 9.37. Usually sheet piling, grouting, de-watering, compressed-air, etc. would be first choices, but sheet piling, for example, is uneconomical for excavations deeper than about 20 m, especially where the wall area exceeds 200 – 300 m². Compressed-air is only viable in heads of water less than 35 m and normally is not considered for depths in excess of 25 m. De-watering and grouting are only effective in a limited range of soil particle sizes. Such constraints

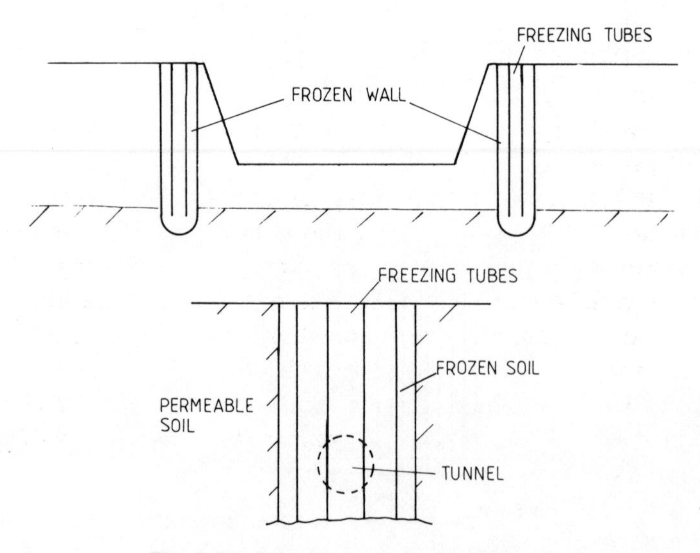

Fig. 9.37 Examples of ground freezing

are not usually imposed on the ground freezing process, but unfortunately the technique is expensive and slow to install.

Methods of ground freezing

The basic principle of ground freezing involves removing heat energy from the surrounding water-bearing soil by passing a refrigerant down the core of a freeze pipe. Isotherms move out from the tube, and over a period of time a column of ice is built up. By spacing the tubes at intervals, the columns of ice can be made to merge to form an enclosing wall (Fig. 9.38).

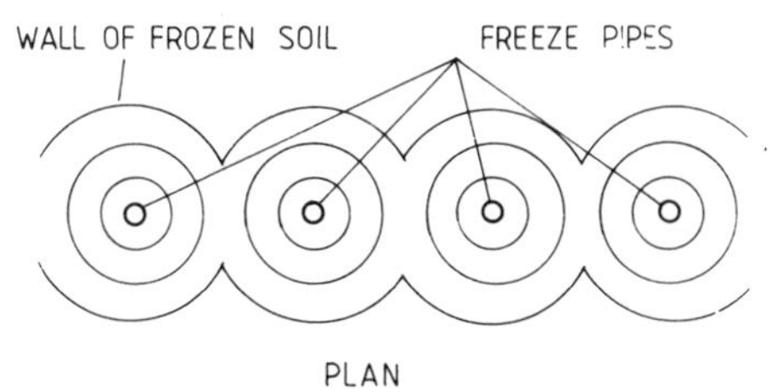

Fig. 9.38 Forming an iced wall

There are basically two freezing methods. (i) the two phase process and (ii) the single phase or direct process.

(i) *The two phase process*

This requires a primary refrigerant such as ammonia or freon to cool a secondary fluid such as glycol or calcium chloride brine. Typical plant is shown in Fig. 9.39. The brine is pumped down the inner core of a double wall pipe at 3 – 5 m/s and returned up the outer core to a chilling vessel.

In the refrigeration plant a gas compressor reduces the volume of the freon gas, followed by cooling and condensing with circulating water. The resulting liquid is then fed through a regulating valve into the chilling coil, where the pressure is released, causing evaporation and cooling of the surrounding circulating brine. The gas is then repressurised in the compressor and the cycle continued.

Brine is a convenient fluid to use down to about −40°C and requires a period of 3 – 10 weeks to complete the freezing process, depending upon the soil type and strength required.

(ii) *The direct process*

This process uses liquid nitrogen which has a boiling point of −196°C.

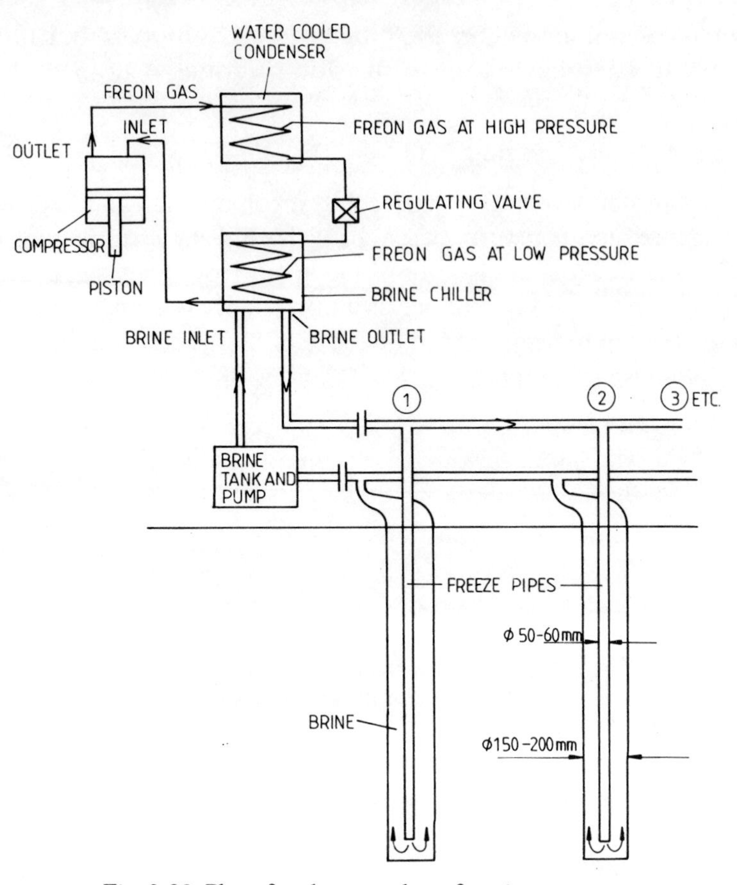

Fig. 9.39 Plant for the two phase freezing process

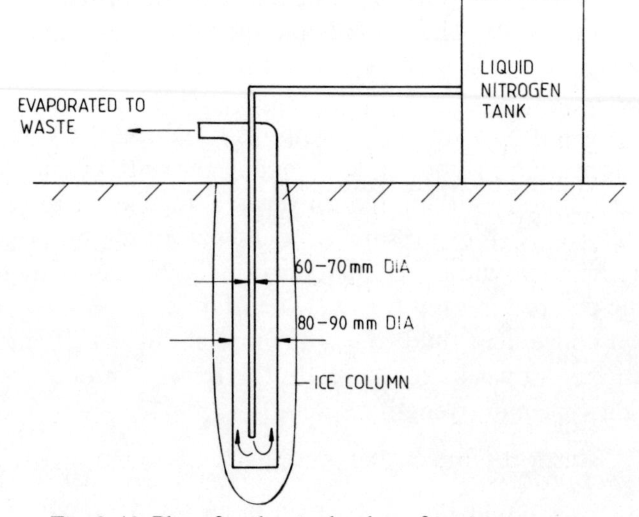

Fig. 9.40 Plant for the single phase freezing process

The fluid is simply passed down the central core of the freeze pipe and evaporated to atmosphere, passing up through the outer core (Fig. 9.40). This process does not involve refrigeration plant, but requires the liquid nitrogen to be stored in an insulated pressure vessel which can be periodically refilled from tanker trucks. Freezing by this method can be up to 50% more expensive than by the brine process, but is much quicker, as an ice wall can often be achieved in a week (approx.).

Ground freezing principles

The freezing process passes through the following stages:

(i) cooling the soil particles and water to 0°C
(ii) removing the latent heat
(iii) reducing the soil and ice to the required temperature.

The heat removed from the ground when reducing the temperature from t_1 to t_2 is governed by the equation

$$Q = \underbrace{[M \times S \times (t_2 - t_1)]}_{\text{Specific heat}} + \underbrace{[M \times L \times (t_2 - 0°\text{C})]}_{\text{Latent heat}}$$

where Q is the heat removed
M is the mass of the body
S is the specific heat of the body
L is the latent heat of ice

Thus for soil and ice

$$Q = V_1 d_s S_s (t_1 - t_2) + V_2 d_w S_w (t_1 - 0°) + V_2 d_i S_i (t_2 - 0°) + V_2 d_w L_i \qquad (9.1)$$

where Q = heat removed in kcal
V_1 = solid volume of soil in m^3
d_s = soil particle density ≏ 2500 kg/m³
V_2 = volume of water in the soil in m^3
d_w = water density = 1000 kg/m³
S_s = specific heat of soil particles ≏ 0.2 kcal/kg—°C
S_w = specific heat of water ≏ 1 kcal/kg—°C
L_i = latent heat of ice = 80 kcal/kg
d_i = density of ice = 900 kg/m³
S_i = specific heat of ice ≏ 0.5 kcal/kg—°C
t_2 = final temperature of ice and soil in °C
t_1 = starting temperature (ambient) of water and soil in °C

Note: V = total volume of loose soil
n = soil porosity
$V_1 = (1-n)V$
$V_2 = V_n$

Assuming that the process is only 50 to 70% efficient (because of heat losses) the size of refrigeration plant required is about 1.5 to 2.0 × Q.

From the above equation, the value of Q needed to freeze 1 m^3 of earth is approximately 40 000 – 60 000 kcal, depending upon the difference between ambient and final temperature.

Theoretical freezing time

$$t = \frac{Q}{hA}$$

where Q is calculated from eqn. (9.1) in kcal
h = heat removed by the freeze pipe, e.g. 200 mm diameter pipe removes 300 – 500 kcal/h—m^2 depending upon temperature
A = surface area of pipe surrounded by soil in m^2.

According to Newton's Law of Cooling, the quantity of heat passing from a cooling liquid per unit of time is proportional to the temperature difference between the liquid and its surroundings (Fig. 9.41). Thus at the early stages the temperature difference between the soil/water and freeze pipe is large and the heat extraction will be high. Near the final temperature heat extraction will be relatively low. In practice, roughly only 30% of the initial rate of heat extraction is necessary to maintain the 'ice' column at its final temperature.

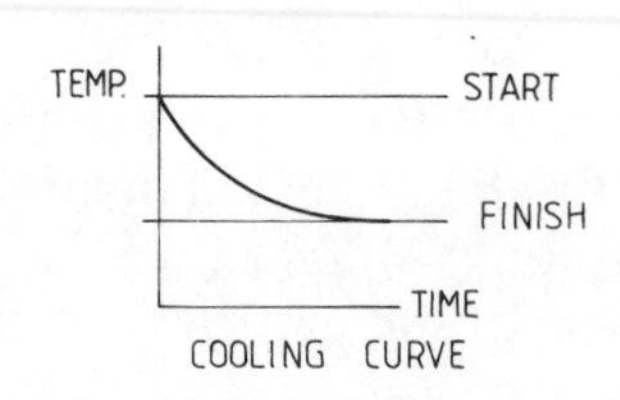

Fig. 9.41 Cooling plotted against time

Refrigeration plant rating

Refrigeration plant is rated in terms of the heat transfer rate for a particular differential between ambient and chilled brine. The unit chosen is usually tons of refrigeration (T_R)

whereby $1\ T_R = 3.517\,\text{kW}$ or $3.517 \times 4.187\ \text{kcal/s}$
or $1\ T_R \triangleq 12000\ \text{Btu/h}$

Plant used for ground freezing typically ranges from 50 - 200 tons rating.
(*Note:* 1 ton = 1.016 Tonnes.)

Disadvantages encountered with ground freezing

(i) Salt water in the soil can make freezing difficult.
(ii) Flowing water causes heat drain, e.g. water movement exceeding 1 - 2 m per day may render ground freezing impracticable. Thus tidal conditions, or adjacent de-watering may pose difficulties.
(iii) Water expands on freezing, but in free draining soil this generally only displaces the surrounding water as the ice front advances and little ground heave results. However, where water is contained in fine gravel soils, ice lenses may develop and so cause heave (Fig. 9.42).

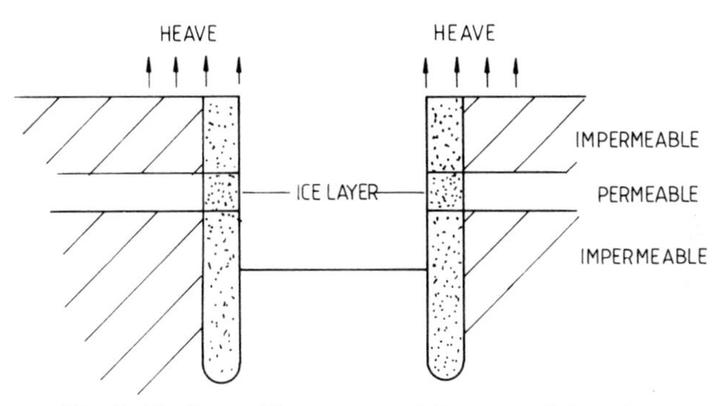

Fig. 9.42 Ground heave caused by ground freezing

Strength of frozen soil

The minimum 'ice-wall' thickness needed to resist bending and shear can be calculated by consideration of the active and passive forces to which the wall is subjected.

Approximate values of compressive strengths attainable in frozen soils are given in Table 9.5.

Table 9.5 Compressive strength of frozen soils

Material	Final temperature −10°C	−15°C	−20°C
Ice	1	1.5	2
Saturated clay	5	7	9
Saturated fine sand	8	13	15
Saturated medium sand	8	13	20

Frozen soil strength (N/mm^2)

Anchorages

In order to provide a clear, unpropped working area in a coffer-dam, anchorage of the shoring system to the surrounding ground may be necessary for large and deep basements (Fig. 9.43).

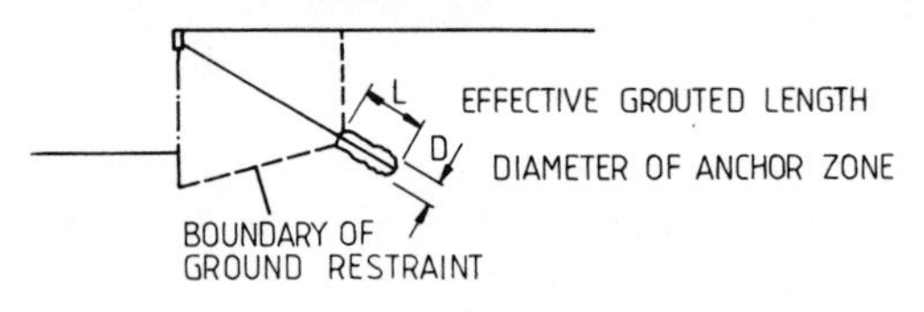

Fig. 9.43 Use of ground anchors

The pull-out force that can be resisted varies markedly depending upon the soil type. The installation methods to deal with the particular ground conditions may be summarised as follows:

Coarse sands and gravels

A hole is bored and lined, using a conventional rotary drilling rig. The steel anchor cable is located in position and cement grout pumped in under pressure. As the lining tube is withdrawn (Fig. 9.44) a bulb of grouted soil is produced around the bottom end of the cable. The resistance to pull-out is therefore highly dependent upon the effective grouted length (L) and the diameter of the anchor zone (D).

Fine to medium sands

Considerably lower pull-out resistance is obtained in these types of soils because the grains are too small to allow cement grout to penetrate the voids. The pull-out force is therefore largely controlled by the diameter of the borehole, although the pressure of the grout may cause some local compaction and so increase the anchor diameter.

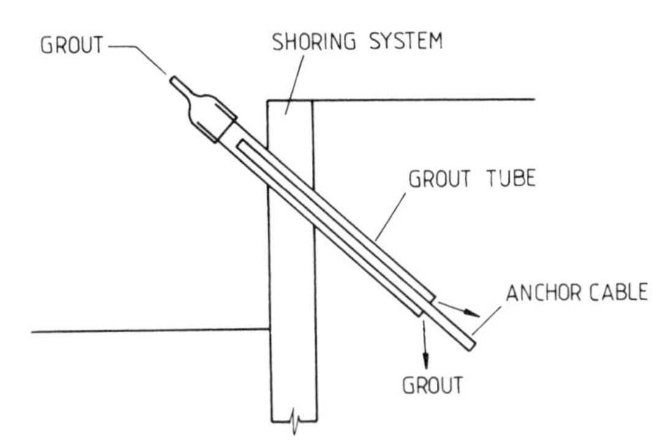

Fig. 9.44 Installing and grouting a ground anchor

Clay

The cohesive strength which can be mobilised between the anchor and clay is fairly small. However, pull-out resistance may be increased by placing gravel in the boring along the anchor length. A lining tube fitted with a non-recoverable point (Fig. 9.45) is used percussively to cause the gravel to penetrate the clay wall of the borehole. The anchor cable is subsequently positioned and grouted in the normal way. Consequently the diameter of the anchored zone is increased.

Other methods of increasing the pull-out resistance involve under-reaming.

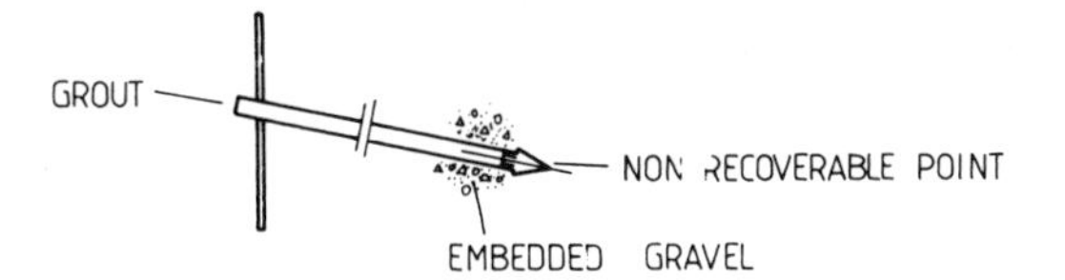

Fig. 9.45 Use of a non-recoverable point grouting method

Rock

The resistance to pull-out is dependent upon mobilising skin friction. A bored hole is drilled and cement grout pumped in to produce the fixed anchor length.

Pull-out loads

These were generated from 100 mm diameter borings and fixed anchor length 4 m. Examples:

Gravel	> 1000 kN
Medium sand	400 kN
Clay	600 kN
Chalk	1000 kN
Rock	> 1000 kN

Note: The load-carrying capacity should always be determined from a test anchor.

Prestressed anchors (*Fig. 9.46*)

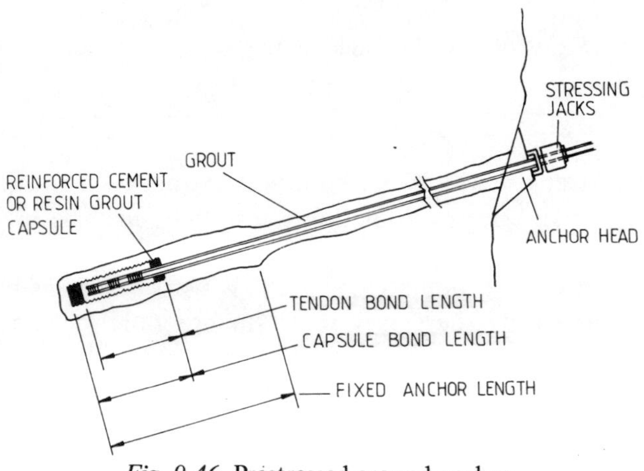

Fig. 9.46 Prestressed ground anchor

Prestressed anchors are mainly used for tying down permanent structures, such as a floating basement, retaining walls subject to imposed loading, etc. and also for improving stability, for example, of a rock face. The design principle aims to minimise movement of the structure when subject to surcharge and service loads.

Borehole diameters (200 – 250 m) are a little larger than conventional anchors. The cable consists of bundles of high tensile steel strands greased and covered by a corrosion-resistant plastic sheath (Fig. 9.47). The fixed anchor length (which has the sheath covering removed) is cast into a 4 – 6 m long corrugated plastic tube, with an epoxy resin. The cable is then positioned in the borehole and grouted in the normal way. The exposed end of the cable is de-sheathed and post-tensioned against a permanent anchorage (Fig. 9.48).

Tension force (example)

Diameter	225 mm
Fixed anchor length	6 m

	gravel	*clay*	*rock*
Tension force (kN)	600	250	> 600

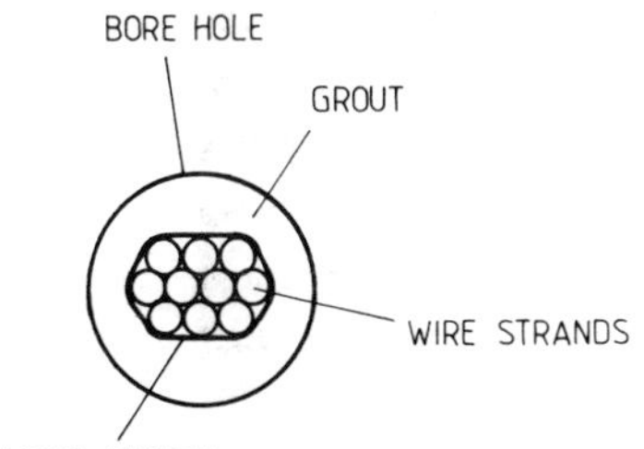

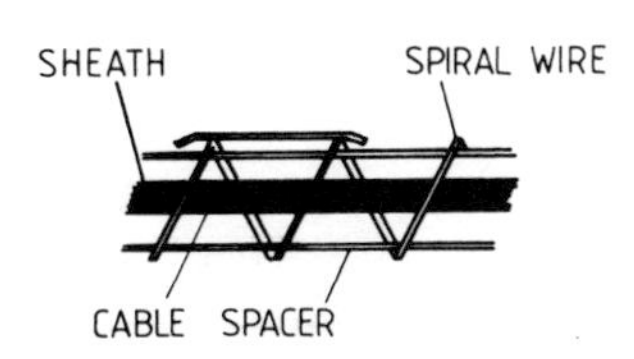

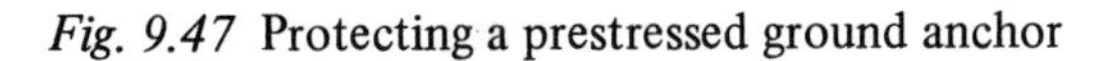
Fig. 9.47 Protecting a prestressed ground anchor

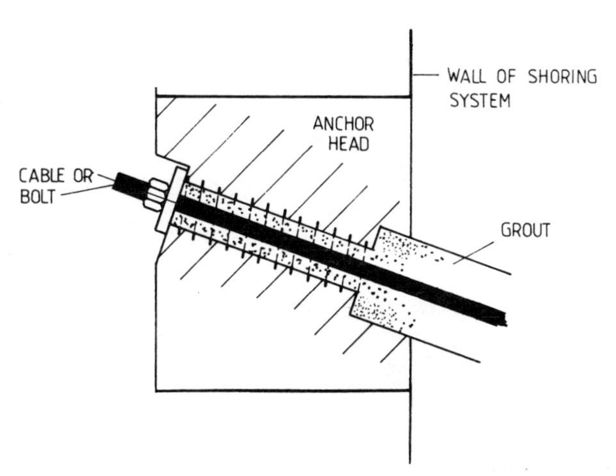

Fig. 9.48 Anchorage for a prestressed anchor

Loss of Tension force

The tension in the cable may lessen with structural movement of the ground, for example up to 5% tension loss was recorded by Littlejohn over a 10 month period for anchors in London clay.

Factors of safety (also see BS.DD81)

Working tension in cable	– 0.6 of breaking tension
Bond between anchor and soil	– 2 for temporary works
	3 for permanent works
Test load	– 1.3 × working or design load

Double-walled sheet piled dam

Construction works to river banks, beaches, docks, etc. require some form of dam to provide temporary protection. Where space permits, a single row of sheet piles or similar supported by an earth embankment (Fig. 9.49) may suffice; in other situations a double row of conventional sheet piles, cross-tied, may be needed (Fig. 9.50).

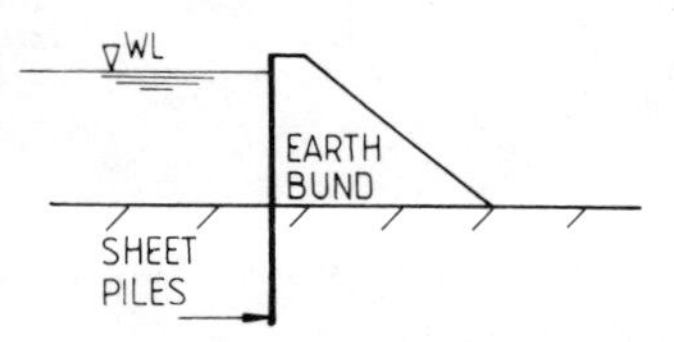

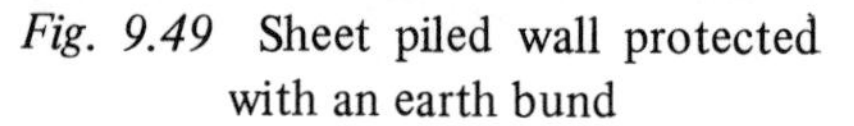

Fig. 9.49 Sheet piled wall protected with an earth bund

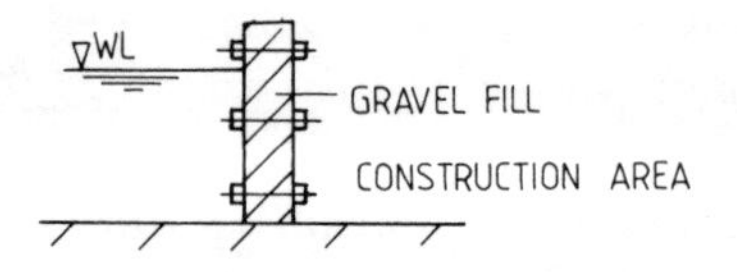

Fig. 9.50 Double-walled sheet piled dam

Design principles

The stability of double-walled structures may be designed on a similar principle to a gravity dam as illustrated in Fig. 9.51.

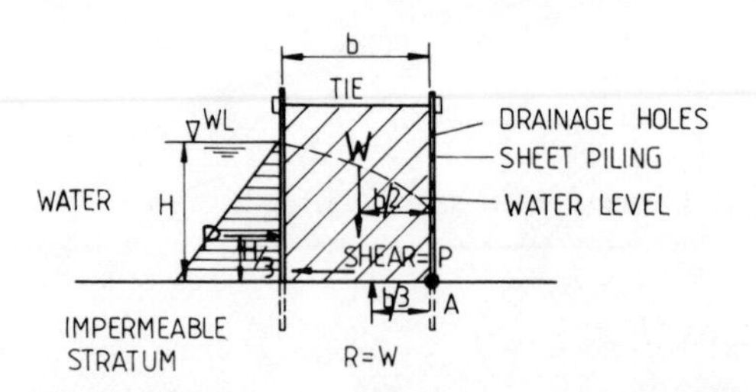

Fig. 9.51 Design principle of double-walled

Thus

$$P = \frac{\gamma_w \times H^2}{2} \qquad \gamma_w = \text{unit weight of water}$$

$$W = \gamma_s \times b \times H \qquad \gamma_s = \text{unit weight of fill/water}$$

Taking moments about A

$$\frac{Wb}{2} = \frac{PH}{3} \times \text{factor of safety } (S)$$

In order to keep the resultant forces P and W within the middle $\frac{1}{3}$ of the base, a ratio of width (b) to height (H) for water-retaining structures of about 0.85 is required.

The structure must also be adequately designed to resist sliding, pile rise and vertical shear. This is beyond the scope of this book, but is adequately treated in references Belz (1970), Terzaghi (1945), Schroeder (1979) and Lacroix, Esrig and Luschev (1970).

Construction arrangement

The dam is filled with permeable material such as sand, gravel or hardcore and adequate drain holes may be provided at various levels in the inner wall to reduce internal hydrostatic pressure. However, if excessive seepage under the dam causes piping then an impermeable bund should be placed in front of the inner wall to increase the seepage path (Fig. 9.52).

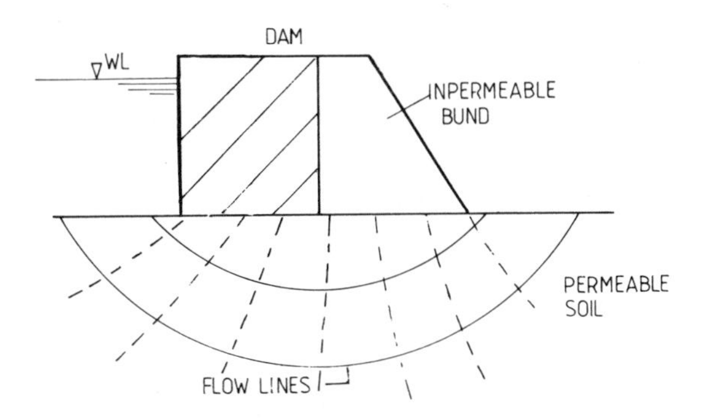

Fig. 9.52 Increasing the seepage path using an earth bund

Cellular coffer-dams (*Fig. 9.53*)

Like the double-walled dam, the cellular structure may be designed as a gravity dam but the need for ties is avoided by taking the active pressure of the fill as ring tension in the piles. Special straight web steel piles are used, but great care is required during pile-driving to avoid straining the clutch, which could subsequently fail when placing the fill, causing complete collapse of the dam (Fig. 9.54). Seepage under the dam may be reduced by driving the piles deeper (see Chapter 6).

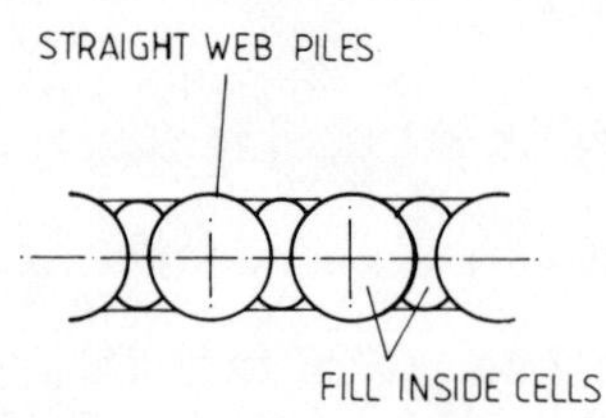

Fig. 9.53 Cellular coffer-dam

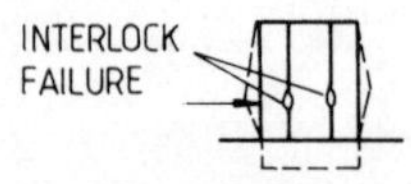

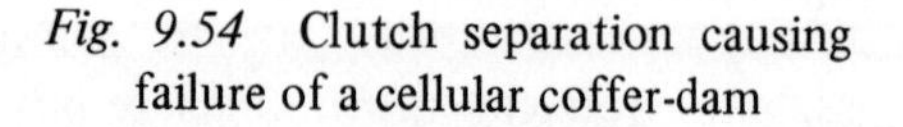

Fig. 9.54 Clutch separation causing failure of a cellular coffer-dam

The hydrostatic pressure inside the cell may be reduced by providing drainage holes on the inside of the coffer-dam (see Fig. 9.51).

Cells founded on clay soil are generally provided with a heavy bund on the inside face to improve stability and in particular to prevent sliding.

Method of construction (*Fig. 9.55*)

A stable position, such as rail track, extended over completed cells, is necessary to facilitate accurate pile-driving. A full ring of straight web piles is pitched and driven around a rigid circular steel frame. If secondary cells are required then the special coupling piles (Fig. 9.56) must be correctly located during the pitching phase, and a further template used to obtain the correct shape for the cell.

A granular fill should be used to aid drainage and shear forces must be removed before any piles are withdrawn in order to avoid a major collapse.

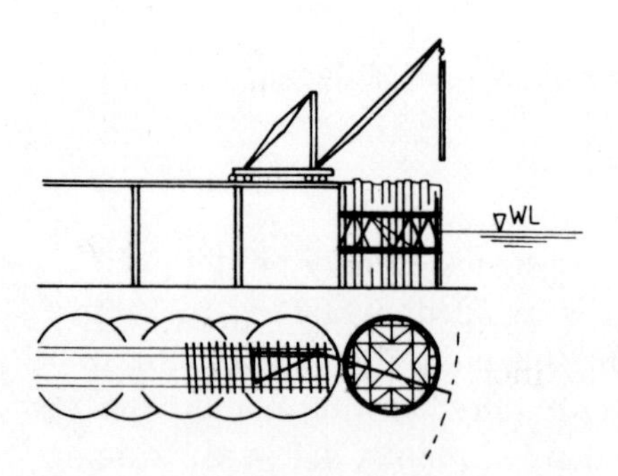

Fig. 9.55 Installing method for a cellular coffer-dam

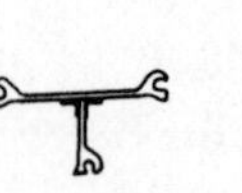

Fig. 9.56 Special straight webbed coupling piles

Open caissons

A caisson constructed in reinforced concrete may be considered as a coffer-dam during the construction phase but it ultimately forms part of the permanent structure – typically a deep foundation or bridge pier.

The caisson consists of an open cell sunk into position by excavating within each cell. It is finally sealed with mass concrete at the base and ballasted to produce an independent foundation, as shown in Fig. 9.57.

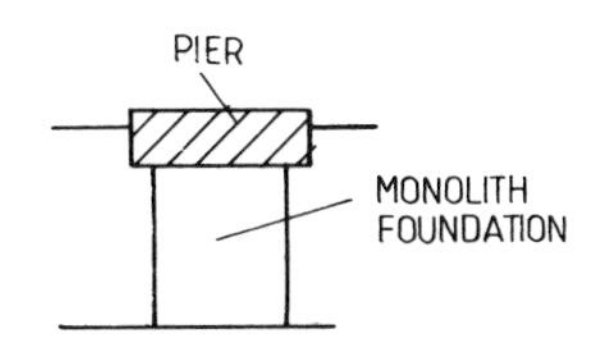

Fig. 9.57 Caisson used to form a foundation or monolith

Applications

Open caissons may be:

(i) used as foundation bases (5 – 10 m deep), where the founding layer of soil or rock would be too hard for sheet piling to penetrate.
(ii) used as coffer-dams or as shoring in highly porous soils which could not be de-watered and excavated in the normal way.
(iii) used as deep foundations (30 – 40 m deep).
They are unsuitable for sinking through soil containing large fragments or where the skin friction between soil and caisson walls is excessive. Also, caissons founded on an uneven or sloping stratum are likely to slip from the final resting place.

Caisson design and construction

(i) The circular caisson offers less surface area for a given base area than a square or rectangular shape and is therefore favoured because of the lower skin friction. However, the shape of the permanent structure incorporating the caisson is often the deciding factor.
(ii) The walls of the caisson should be constructed carefully. Rippled and non-parallel walls are likely to increase sinking resistance.
(iii) Verticality during sinking of the caisson can be controlled by choosing a multi-cell structure so that individual areas may be excavated in isolation (Fig. 9.58).
(iv) The size and shape of the cells should be arranged for maximum

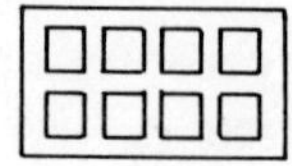

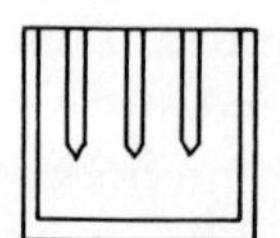

Fig. 9.58 Multi-cell reinforced concrete caisson

open digging. Rectangular cells are preferred, especially in clay soil. There is a tendency to arch and wedge around the cutting edge if a circular form is selected.

(v) The shape of the cutting edge should be varied to suit the soil type. The stiffer the soil, the steeper the angle of the cutting edge, but 35° from the vertical is typical (Fig. 9.59). The cutting edge should also be well reinforced to cope with high stress and therefore a steel shoe at the base is commonly incorporated.

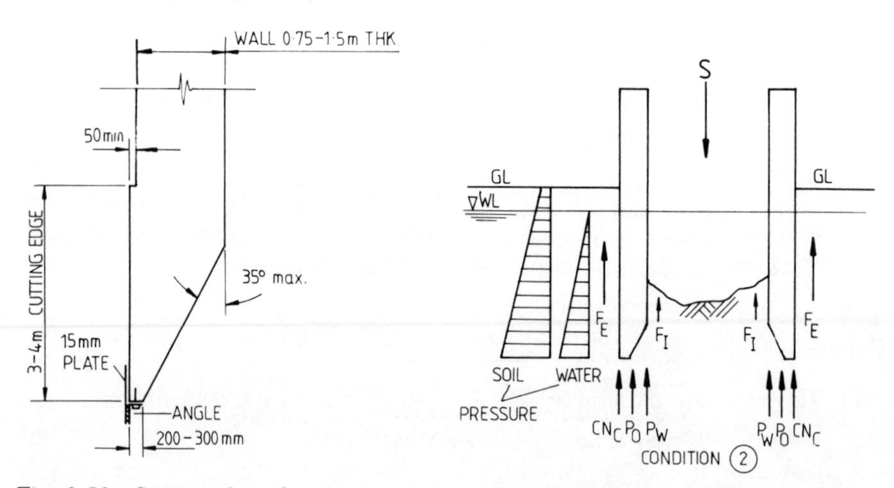

Fig. 9.59 Cutting shoe for a caisson

Fig. 9.60 Sinking principles of a caisson

Sinking theory

Consider the section of caisson wall and cutting edge shown in Fig. 9.60.

The caisson comes to rest when the sinking force is overcome by skin friction and ultimate bearing resistance.

Working in total stresses,

$$S = F_E + F_I + (CNc + P_o + P_w)A$$

where S = total weight of caisson and kentledge
F_E = external skin friction
F_I = internal skin friction
CN_c = net ultimate bearing resistance
P_o = effective overburden pressure
P_w = water pressure
A represents the area of monolith shoe.
CN_c is part of Terzaghi's equation for net ultimate bearing resistance which is true for $\phi = 0°$ (ϕ is the internal angle of friction for the soil).
C = cohesive strength of the soil
N_c = average value of 7

The value of C used in calculating F_E and F_I should be the remoulded shear strength, in order to reflect the effects of soil movement as the caisson sinks. However, practice has demonstrated that calculation of skin friction based on laboratory soil tests is unreliable. Terzaghi and Peck, however, have recorded loads needed to free stuck caissons and these have been interpreted to give values for skin friction as shown in Table 9.6.

Table 9.6 Values of skin friction for caissons

Type of soil	Skin friction kN/m²
Silt and soft clay	7 - 28
Very stiff clay	50 - 200
Loose sand	12 - 35
Dense sand	35 - 67
Dense gravel	50 - 100

From K. Terzaghi and R.B. Peck (1967). *Soil Mechanics in Engineering Practice.* John Wiley & Sons, Inc. New York.

Calculations should be made for several conditions, as shown in Fig. 9.61, including the initial settlement, intermediate points and when the caisson is constructed to its full height. In this manner the amount of kentledge, height of caisson above ground level, depth of undercutting, etc. can be determined for controlling the amount of sinking.

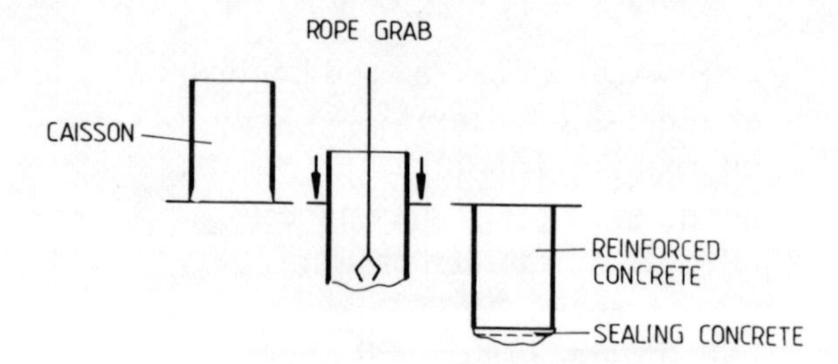

Fig. 9.61 Sinking procedure for a caisson

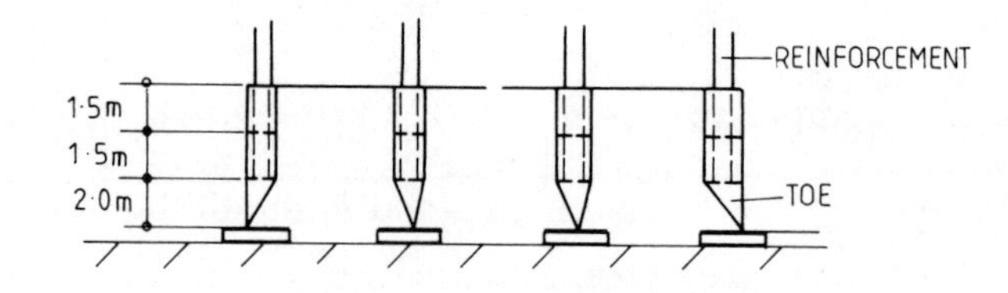

1	1	1	1	1	1
1	2	2	2	2	2
1	2	3	3	2	1
1	2	3	3	2	1
1	2	2	2	2	1
1	1	1	1	1	1

GRABBING SEQUENCE

Fig. 9.62 Excavating sequence for a multi-cell caisson

Sinking and construction procedure (*Fig. 9.62*)

(i) Prepare a level bed and establish concrete pads where the toe of the walls is to be constructed.

(ii) Construct the toe and walls up to about 5 m.

(iii) Break out the concrete pads and allow the caisson to settle. The amount of initial sinking will depend upon the strength of the soil.

(iv) Grab within the cells – this is the most commonly used method of excavating from the open cells. Grabbing must be done in a logical sequence to ensure that verticality of sinking is maintained. The grabs can either be mounted upon derricks or on mobile cranes. Sometimes other methods are appropriate, e.g. in sands, a jetting pipe often helps to loosen the soil particles, which are then lifted to the surface by an air-lift pump. This equipment (Fig. 9.63)

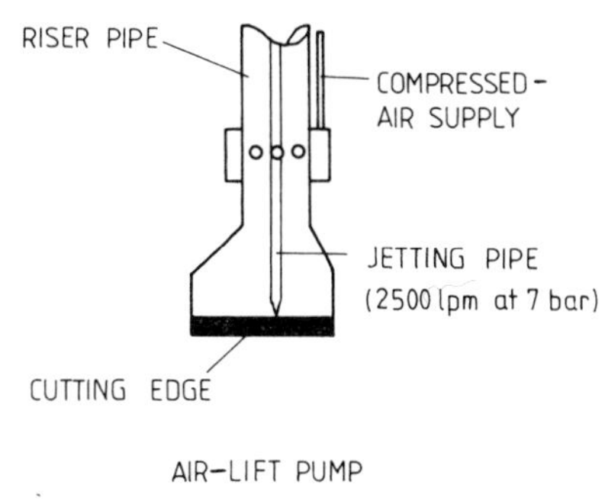

Fig. 9.63 Air-lift pump

contains compressed-air which is injected into a riser pipe, and the aerated water, being of lighter density, flows up the riser pipe taking the loosened soil with it.

(v) The excavation is continued until about 1 lift of concrete (1.5 m) freeboard is remaining. Then the walls are extended. This in itself may result in significant sinking if the soil is weak. The height of the walls should be such that construction access is easy and the centre of gravity does not become too high to jeopardise the accuracy of sinking. About 6 - 7 m of freeboard is usual. Excavating within the cells is then continued.

(vi) On reaching founding level, open caissons are sealed (Fig. 9.64) by depositing a layer of concrete under water in the bottom of the wells (plugging). The wells are then pumped dry and further concrete or filling material is added to give dead weight (hearting).

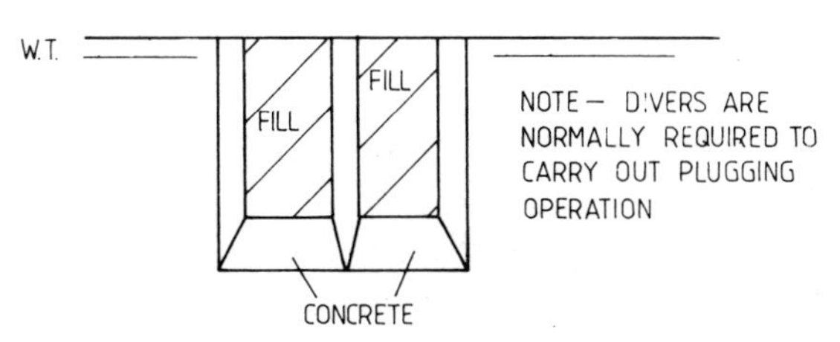

Fig. 9.64 Plugging and hearting for a caisson

Control of verticality

(i) add concrete (kentledge on one side or the other)
(ii) jet under the cutting edge on the 'hanging' side
(iii) control the sequence of cell excavation

Releasing a 'hanging' caisson

Sometimes a caisson will meet a lens of material with a high friction coefficient, such that continued excavation in the cells causes no further sinking. If jetting or adding kentledge does not free the structure, then the only recourse is to reduce the natural level of water inside the cells (Fig. 9.65). The result is a reduction in the uplift pressure, thus,

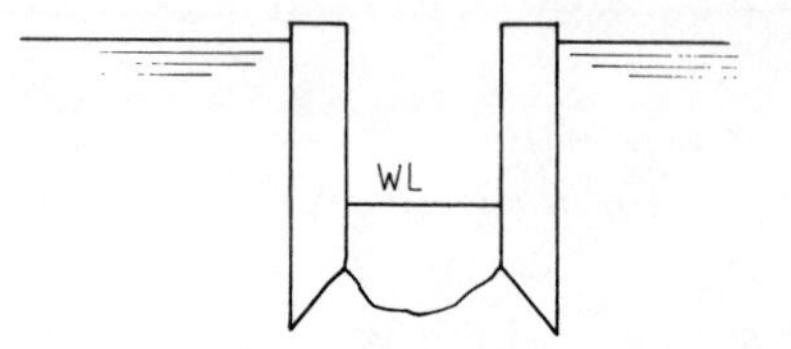

Fig. 9.65 Freeing a hanging caisson

in effect, increasing the sinking force. In fine grained soils, however, caution should be exercised to avoid the possibility of a 'blow', i.e. piping, (see page 108) occurring, causing the caisson to settle and possibly tilt quite suddenly and significantly.

Sinking aids

Introducing water, compressed-air or bentonite slurry can reduce skin friction considerably. The material is introduced through pipes cast into the walls as shown in Fig. 9.66.

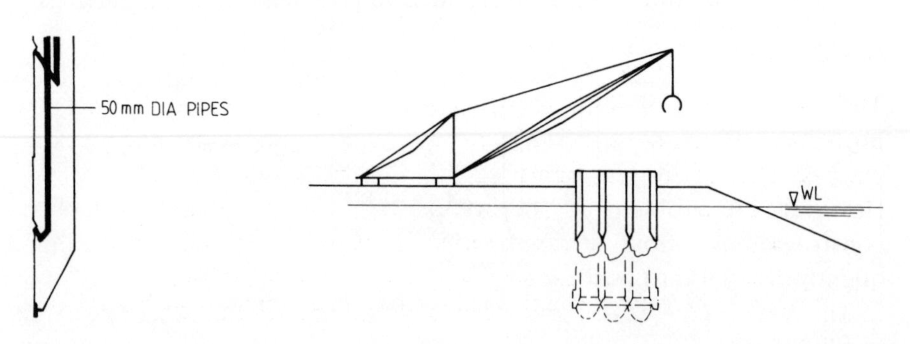

Fig. 9.66 Aids to sinking

Fig. 9.67 Grabbing method suitable for caisson sinking

Output

Excavation is usually carried out from a rope-suspended grab of approximately 0.75 – 1.5 m^3 struck capacity (Fig. 9.67). The operator is working blind and digging through water and so normal output rates cannot be expected.

Examples for 0.75 m³ grabbing bucket

Easy dig - 10 m^3/h
Medium dig - 6 m^3/h
Hard dig - 4 m^3/h

Note: These are bulked volumes and do not allow for stoppages or 'stuck' periods.

Constructing a caisson from islands

Caissons for bridge piers and docks or lock walls are generally started from temporary islands as shown in Fig. 9.68. However, when deep water is involved a closed or box caisson is usually necessary.

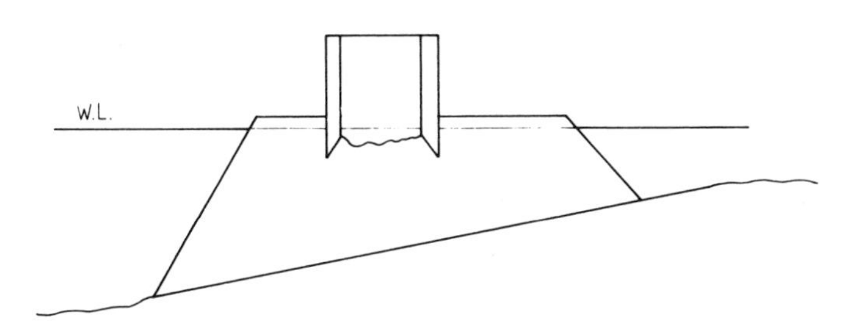

Fig. 9.68 Constructing a caisson from an island

Box Caissons

The use of open caissons for the construction of dock walls, harbour protection, breakers, watch towers, etc. is often impractical or uneconomic because of the temporary works necessary, e.g. islands or falsework. However, the problem can be avoided with the box caisson, which is constructed as a floating vessel on dry land or in a dry dock and subsequently floated into position and sunk onto its foundation.

The vessel can be of any desired shape, but a full structural and stability analysis must be undertaken at the launching, transporting and sinking phases, in order that planned and controlled procedures can take place.

Foundations for box caissons

A gravel or crushed rock bed provides a suitable foundation if erosive

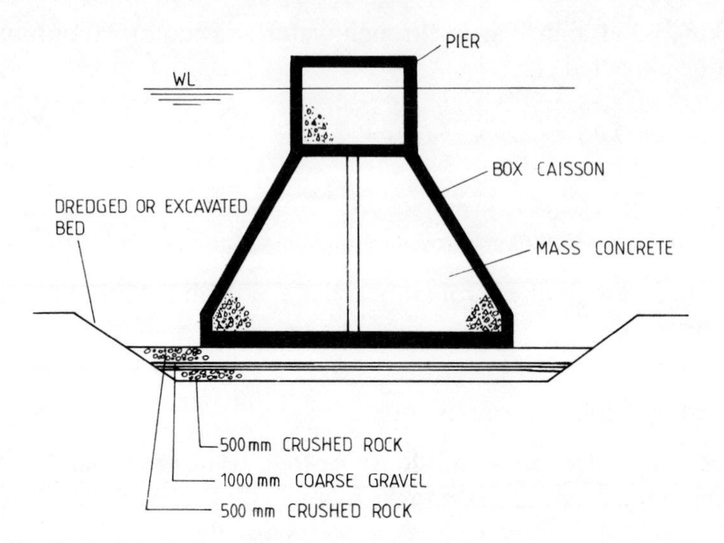

Fig. 9.69 Caisson foundation where water currents are not excessive

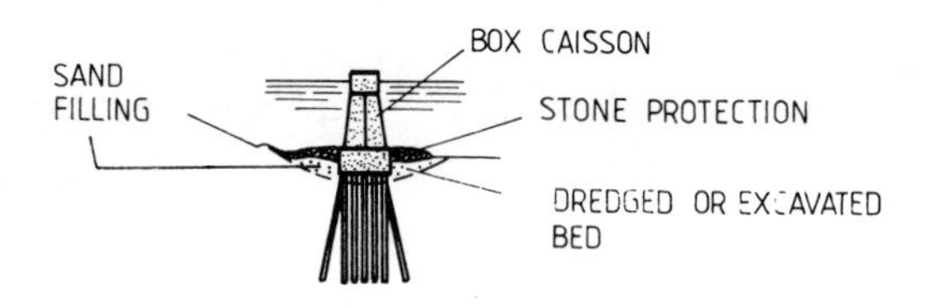

Fig. 9.70 Piled caisson foundation in rapid flowing water

currents are not present (Fig. 9.69). In fast flowing waters, such as rivers, a piled foundation is more appropriate (Fig. 9.70).

Pneumatic caissons (*see also compressed air working in tunnels, Chapter 10*)

A pneumatic caisson is a box made from steel or concrete constructed like a diving bell at its base. Compressed air is pumped into the bottom section, thereby excluding water, to allow workmen to excavate in a dry chamber. The structure is sunk in a similar fashion to the open caisson (Fig. 9.71) but men and materials must be passed through an air locking system (Fig. 9.72).

The pneumatic caisson is necessary where a foundation in open water must be founded well below the river or sea bed, or alternatively on dry land, where free-flowing soils might cause settlement of adjoining buildings, if an open caisson were chosen.

However, because men have to work in compressed air, the relatively

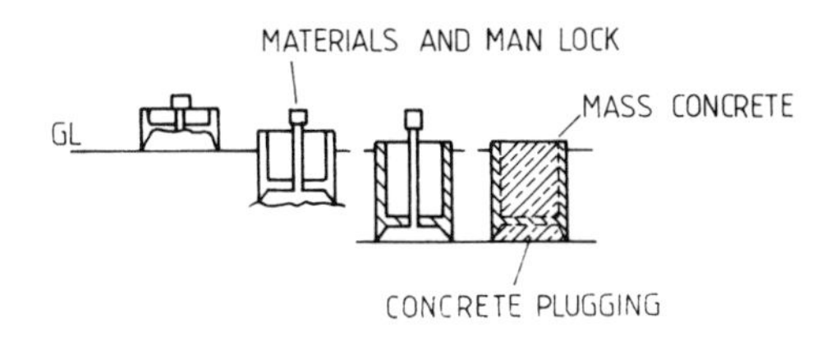

Fig. 9.71 Sinking procedure for a pneumatic caisson

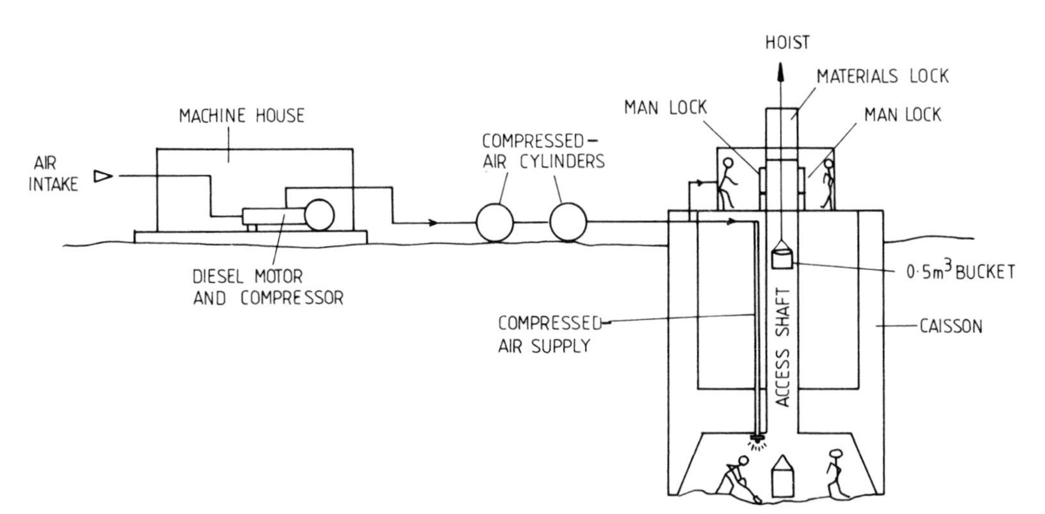

Fig. 9.72 Air locking system for a pneumatic caisson

short work shifts and high wages make the method expensive and, today, it is generally used only as a last resort.

Sinking theory

The calculations required to determine stability during sinking and when in the final resting position are similar to those required for the open caisson, with the addition of the extra uplifting force in the compressed chamber (Fig. 9.73).

Safety regulations for working in compressed-air limit pressures to about 3.5 bar, which is equivalent to about 35 m depth in water.

Sinking control

(i) Excavate symmetrically.

(ii) Maintain a berm around the inside cutting edge to reduce air leakage.

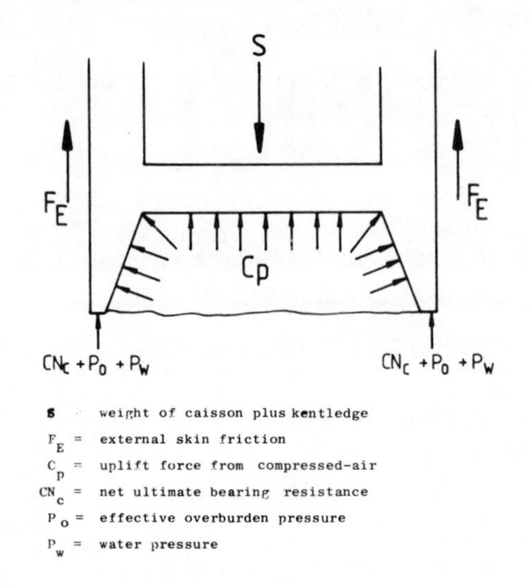

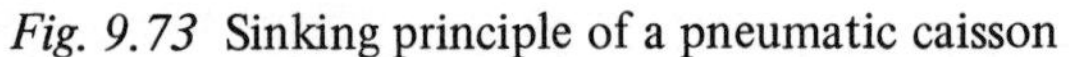

Fig. 9.73 Sinking principle of a pneumatic caisson

(iii) (a) In sands the inside pressure should be about 0.1 bar – higher than that required to maintain dry working conditions. Too much pressure may cause a sudden blowout and subsequent loss of pressure, causing material to flow into the chamber. In these circumstances, remove the men, flood the chamber and then re-pressurise.

(b) In clays there is less air leakage, but attention must be given to the possibility of a sudden pressure drop when a layer of sand or gravel is encountered.

(iv) Excavate continuously wherever possible to allow continuous movement and avoid a build-up of static friction.

(v) Have standby compressed-air plant and decompression facilities available.

(vi) The regulations governing work in compressed-air are dealt with fully in the references and guidelines given in Chapter 10.

Output

An excavation rate by hand in easy material would be about 0.5 m^3 per man hour. Up to ten men in the digging gang is common, working the hours governed by the working in compressed-air regulations.

Typically, sinking rates vary between 0.05 and 0.3 m per day, depending upon soil type.

A topside gang is required to operate the compressor, hoist and lock and in addition it is common practice to have a fitter and a foreman.

Bibliography

Appolonia, D.J. (1980) 'Soil-bentonite slurry trench cut-offs' J. Geotech. Eng. Div., ASCE, April.

Belz, C.A. (1970) 'Cellular structure design methods and installation techniques'. *Proc. Conf. on design and installation of pile foundations and cellular structures.* Envo Publishing Co., Lehigh Valley, USA.

Boyes, R.G.H. (1975) *Structural and Cut-off Diaphragm Wall Construction.* Applied Science Publishers; London.

BS6031 (1981) (formerly CP2003) *Earthworks.* British Standards Institution.

BS1377 (1972) *Methods of Testing Soils for Civil Engineering Purposes.* British Standards Institution.

BS5930 (1981) (formerly CP2001) *Site Investigations.* British Standards Institution.

B.S.P. International Ltd., Sheet Piling Systems Pocketbook, Ipswich.

CP2004 (1972) *Foundations.* British Standards Institution.

Dawson Ltd., Sheet pile design and installation. Luton.

DD81 Draft Code for ground anchors (1982) British Standards Institution.

Fages, R. and Gallet, M. (1973) 'Calculations for steel piled or cast in situ diaphragm walls'. *Civil Engineering and Public Works Review*, **68** (809), Dec.

Frydman, S., Melnik, J., & Baker, R. (1979) 'Effect of prefreezing on the strength and deformation properties of granular soils'. *Soil Foundation*, **19k** (4), Dec.

Grant, W.J. 'Nitrogen applications – ground freezing'. BOC report 3949.

Haiun, G. (1975) 'Recent progress in the efficiency of steel sheet piling techniques'. *Travaux*, **478**, Jan. (French).

Hanna, T.H. (1980) Design and construction of ground anchors. CIRIA Report 65.

Harris, J.S. (1972) 'The control of ground water by freezing'. *Proc. Inst. Civil Engineers*, Feb.

Hiremech Plant Ltd., Hydraulic Shoring Systems. York.

Icos Ltd. (1983) Structural diaphragm walls. Buckingham Gate, London.

Irvine, D.J., & Smith, R.J.H. (1983), Trenching practice, CIRIA Report 97.

Jessberger, H.L. (1980) 'State of the art report – ground freezing'. Int. Symp. 2nd, ISGF 80, Norwegian Institute of Technology, Trondheim, June.

Krings Company, Trench Linings, Clevedon, Avon.

Lacroix, Y., Esrig, M.B. & Luschev, U. (1970) Design, construction and performance of cellular coffer-dams', *ASCE Speciality Conference*, Cornell Univ.

Littlejohn, G.S. (1968) 'Recent developments in ground anchor construction'. *Ground Engineering*, **1** (3).

Littlejohn, G.S. (1980) Design Estimate of Ultimate load-holding capacity. Ground anchors. *Ground Engineering*, **13**, Nov. ICE. London.

Lorenz/Fehlmann method of caisson sinking (1967) *Water and Water Engineering*, **71**, April.

Mabey Ltd., Trench boxes. Twyford, Reading.

MacKay, E.B. (1979 and 1982) 'Proprietary trench support systems'. CIRIA Technical Note 95.

MacKay, E.B. (1981) 'Guide to trench support methods to 6 m depth'. CIRIA.

Maitland, J.K. et al (1979) Model study of cellular sheet pile cells. *J. Geotech.* Eng. Div., ASCE, **105** (7), July.

Miles, M.M. & Boyes, R.G.H. (1982) 'Slurry trenching development'. *Civil Engineering*, April. Morgan Grampian Ltd.

Oosten, P.A. (1974) 'Computer analysis of a sheet pile wall', *Polytechnisch Tijdschorft*, **29** (10), May (Holland).

Ridout, G. (1981) 'Steel sheet piling – 50 years of driving'. *Contract Journal*, **303**, October.

Rojo, J.L. (1980) 'Borne substitute liquids for soil freezing at very low temperatures'. Ground Freezing, Int. Symp. ISGF '80, Norwegian Inst. of Tech., Trondheim, June.

Schroeder, W.L. and Maitland, J.K. (1979) 'Cellular bulkheads and coffer-dams'. *J. Geotech.* Eng. Div., ASCE, **105** (7), July.

Scott, C.R. (1980) *Soil Mechanics and Foundations.* Applied Science Publishers, London.

Shorco Ltd., Drag boxes. Churwell, Leeds.

Short, N.R. and Harris, F.C. (1983) 'A comparison between trench support systems and traditional methods of shoring' *Building Technology and Management.* Feb.

Speedshore Ltd., Trench excavation and shoring systems. Wetherby.

Spett, C. (1979) 'Anchors: performance prediction'. *Consulting Engineer*, **43** (12), Dec.

Stoss, K. & Valk, J. (1979) 'Uses and limitations of ground freezing with liquid nitrogen'. *Eng. Geology*, **13**, (1–4), Apr. Bochum, Germany.

Terzaghi, K. & Peck, R.B. (1967) *Soil Mechanics in Engineering Practice.* Wiley, N.Y.

Terzaghi, K. (1945) Stability and stiffness of cellular coffer-dams. Transactions, ASCE, **110**.

Tomlinson, M.J. (1969) *Foundation Design and Construction*, Pitman, London.

Tschebotarioff, G.P. (1973) *Foundations, Retaining and Earth Structures.* McGraw-Hill, N.Y.

Veranneman, G. & Rebman, D. (1979) 'Ground consolidation with liquid nitrogen'. *Eng. Geology*, **13**, (1–4), Apr., Bochum, Germany.

Weltman, A. (1977) 'Stating the case for bentonite – boring for piles'. *Contract Journal*, **279**, Sept.

Winterkorn, H.F. & Fang, H.Y. (1975) *Foundation Engineering Handbook.* Van Nostrand Reinhold, N.Y.

Yang, N.C. (1965) 'Condition of large caissons during construction'. *Highways Res. Rc.* **No. 74**, Washington DC.

CHAPTER TEN

Tunnelling

Tunnels are used for underground road and rail transport, for hydro and sewage conduits, drainage and many other services.

The methods adopted for the construction of tunnels vary from hand excavation to sophisticated power-driven tunnelling machines, depending upon the soil type, ground conditions, length of tunnel, etc. The techniques, however, may be roughly classified into five groupings:

(i) tunnelling in rock
(ii) tunnelling with shields
(iii) tunnelling with machines
(iv) open-cut tunnelling
(v) small bore tunnelling

Tunnelling in rock

In short lengths of tunnel or where the diameter or shape is unsuitable for excavating machines, the traditional use of explosives to advance the heading is often favoured. Depending upon the stability of the rock the heading can be tackled by one of the following methods:

(i) advance the heading in full face without lining
(ii) advance the heading in drifts followed by full face lining
(iii) advance the heading in drifts followed by lining progressively in stages

(ii) and (iii): classical methods (rarely used in modern times except for pilot tunnels)

(iv) modern improvements to (ii) and (iii)

Full face heading without lining

Where the rock is stable and self-supporting, excavation of the heading in full face (Fig. 10.1(a)) is suitable for tunnels up to about 200 m^2 in cross-section. The procedure is to drill holes for explosives and advance the heading by blasting. The broken debris is loaded into trucks or

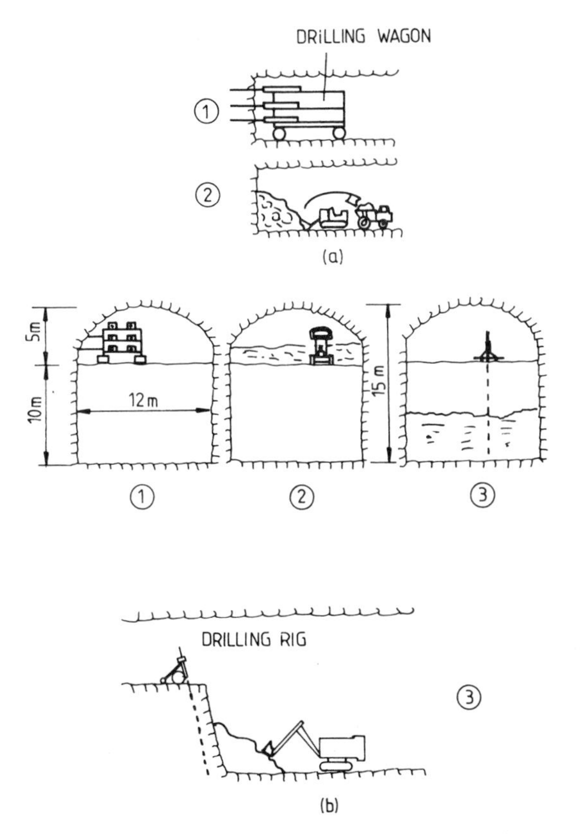

Fig. 10.1 Full face heading in rock without lining

onto a conveyor or transported by rail car away from the heading. Single or multiple drifter drills are commonly used to produce the holes for the explosive and these are discussed in detail in Chapter 3. The techniques of blasting are covered later in this chapter and in Chapter 4.

If the heading is large enough to be sub-divided and allow the use of excavating and materials handling plant and equipment, then progress can be increased by benching, as shown in Fig. 10.1(b).

When the cross-sectional area of the tunnel exceeds approximately 150 - 200 m^2, the tunnel roof and sides will often require propping, and work should then be carried out progressively from a series of sub-headings or drifts.

Classical methods

Advance the heading in drifts followed by full face permanent lining
The heading is usually opened out in segments, working from a small pilot tunnel driven the full length, which provides ventilation and a

means of access for transporting materials and for carrying out temporary works such as grouting. If the rock is fairly stable and unlikely to suffer fracturing and bursting as the stresses in the rock are released, the drifts are sequentially advanced and temporarily supported until a complete ring of permanent lining can be put in place. Generally the English (Figs. 10.2 and 10.3) and Austrian (Fig. 10.4) methods follow this procedure.

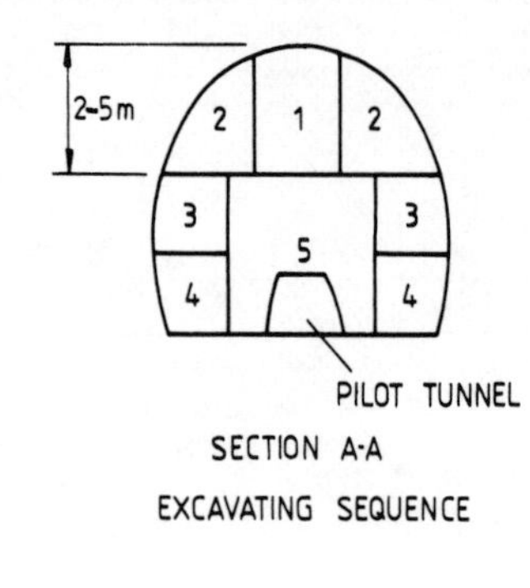

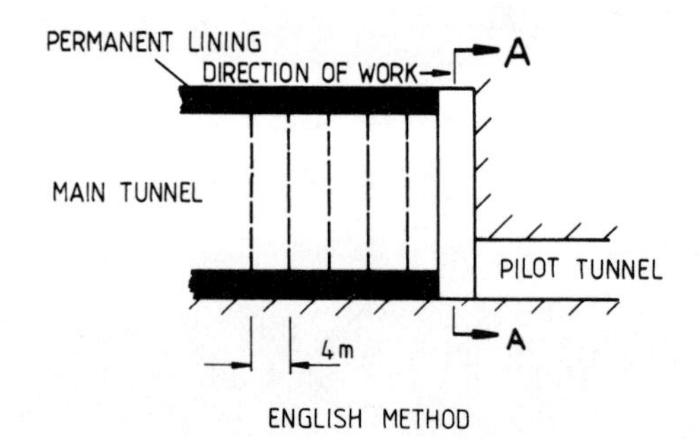

Fig. 10.2 English method

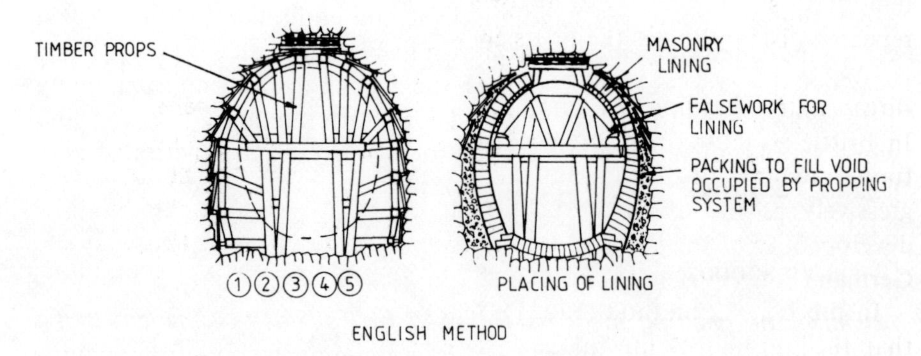

Fig. 10.3 Propping sequence in the English method

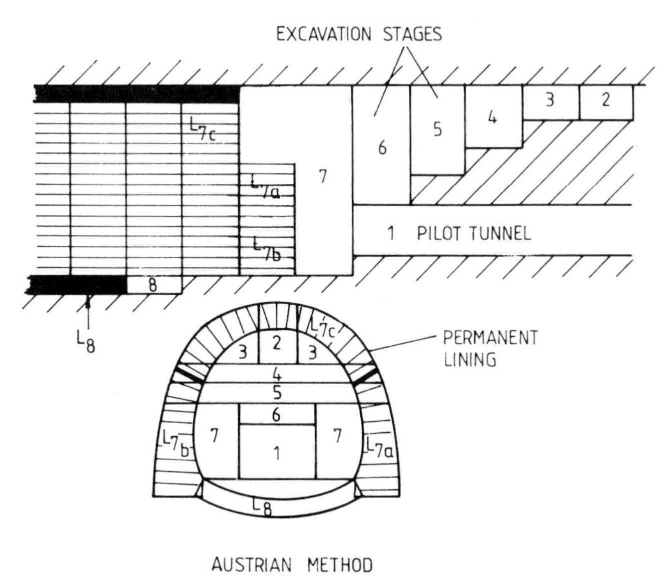

Fig. 10.4 Austrian method

In the English system, the top segment (1) may sometimes be driven the full length to improve ventilation. As the heading is enlarged, temporary support is given by timber props. The permanent lining is constructed when a full face has been excavated and the props are then removed. The face is progressively advanced in this manner.

The Austrian method is similar to the English system in that almost a complete ring of lining is positioned when a full circumference is available, but unlike the English method the segments of the heading are advanced in a series of drifts. As a consequence, the need to prop the full face of the heading (thus causing complicated propping), as in the English method, is avoided. However, the Austrian method requires temporary support of the roof for a longer period, which is subject to repeated dismantling as the drifts are developed.

Advance the heading in drifts followed by staged permanent lining
In brittle, fractured or soft rock, to avoid leaving the full face of the tunnel temporarily propped, the permanent lining is installed progressively as the drifts are advanced. The three classical techniques developed to deal with these conditions are the Belgian (Fig. 10.5), German (Fig. 10.6) and Italian (Fig. 10.7) methods.

In the Belgian method (Fig. 10.8) a sequence of drifts is used, such that the top half of the tunnel face is excavated and propped and the lining is then installed to form a flying arch. The remaining segments are subsequently excavated, working outwards from the centre.

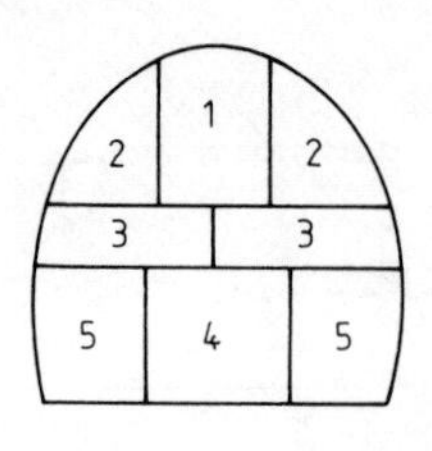

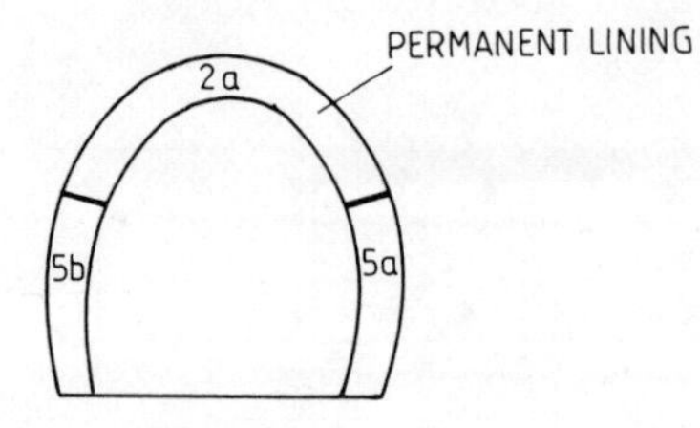

EXCAVATING SEQUENCE LINING SEQUENCE

BELGIAN METHOD

Fig. 10.5 Belgian method

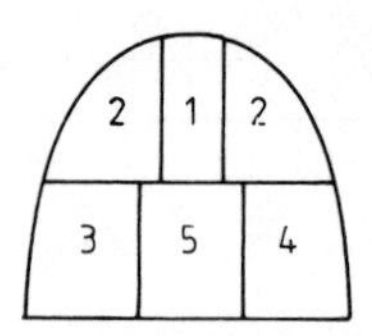

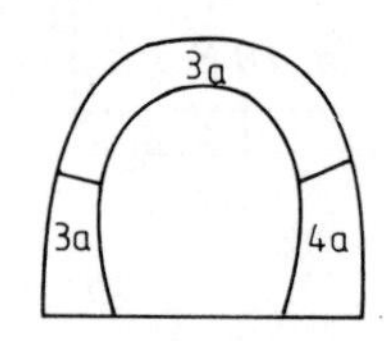

EXCAVATING SEQUENCE LINING ERECTION

GERMAN METHOD

Fig. 10.6 German method

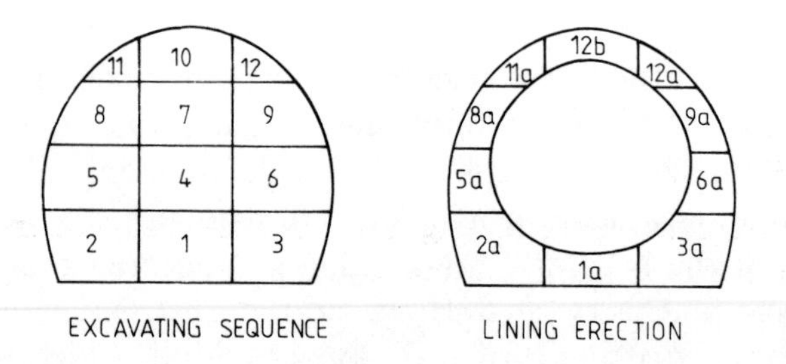

ITALIAN METHOD

Fig. 10.7 Italian method

In less stable rock, the German method is preferred. The core is left until last and provides a stable propping base. The side linings are erected next, followed by the roof lining. Like the Belgian method this sequence is performed in progressive stages.

When tunnelling in very unstable conditions, the Italian method is commonly adopted. Small drifts for subheadings are required, working from the base upwards.

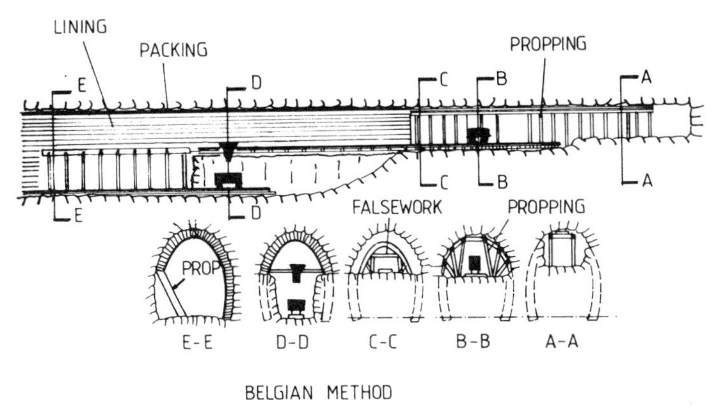

Fig. 10.8 Propping sequence in the Belgian method

Modern improvements to the classical methods

Because of the high cost of timbering and the specialist skills required to install the temporary support, the classical methods of tunnelling have gradually been modified as technical improvements to support systems and permanent linings have been developed. The new temporary support techniques (which can also form part of the permanent lining) are simpler to install and include:

(i) sprayed concrete
(ii) rock bolting
(iii) stiffening ribs and liner plates
(iv) full lining

The New Austrian method (Fig. 10.9) is proving to be an economical sequence in rock-tunnelling when used in association with the new support methods.

(i) *Sprayed concrete*

It can be seen from Table 10.1 that guniting is an appropriate temporary lining and support for virtually all types of rock tunnelling. However, the method is most economical with full face working. In loose rock, additional support using mesh or rock bolts or steel ribs may be required. Sprayed concrete of aggregate size up to 25 mm can be used, and with chemical additions strength, adhesion and setting time can be greatly improved. This type of lining may also serve the purpose of a permanent lining, depending upon the tunnel design.

A method of achieving good results is illustrated in Fig. 10.10.

Table 10.1 Guniting for temporary or permanent lining (Szechy)

Type of rock	Gunite thickness (mm)	Max. unbridged span (m)	Max. unbridged time	Comments
Firm	–	–	Non-specifiable	Support not required
Loosening over time	20 - 30	3 - 4	Up to ½ year	Gunite only required in the arch roofing
Slightly friable	30 - 50	2.5 - 3.5	1 week	Gunite only required in the arch roofing. Temporary propping needed
Friable	50 - 70	1 - 1.5	½ day	Temporary propping followed by mesh reinforced gunite
Very friable	20 - 150	0.5 - 0.8	½ h	Extensive temporary propping followed by mesh reinforced gunite
Immediate light ground pressure	150 - 200	Less than 0.5	1 - 2 min	Support from steel ribs, followed by mesh reinforced gunite
Immediate heavy ground pressure	–	0.1 - 0.2	A few seconds	Guniting not suitable

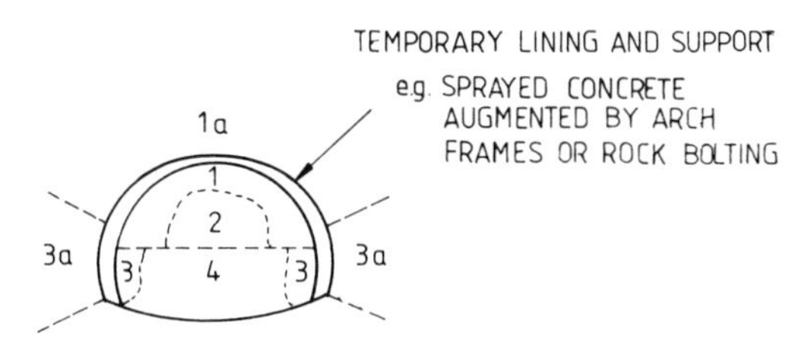

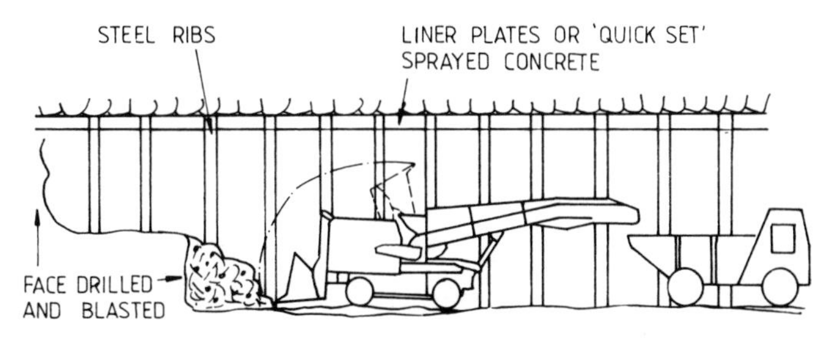

Fig. 10.9 New Austrian method

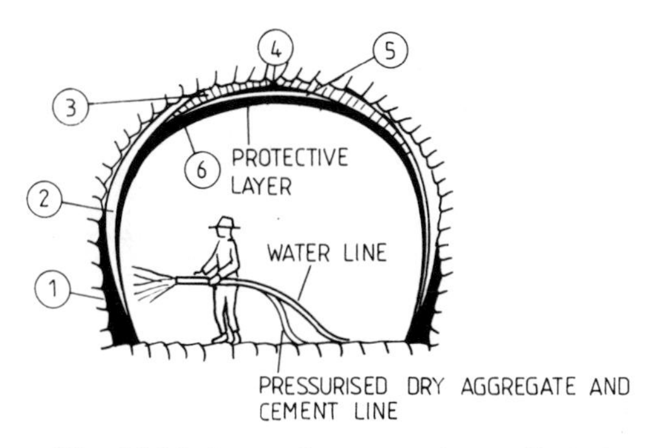

Fig. 10.10 Sprayed concrete lining (Leins)

(ii) *Rock bolting*

Rock bolting, often applied in combination with sprayed concrete, can be used as a temporary or permanent support system as indicated in Fig. 10.11 and Table 10.2.

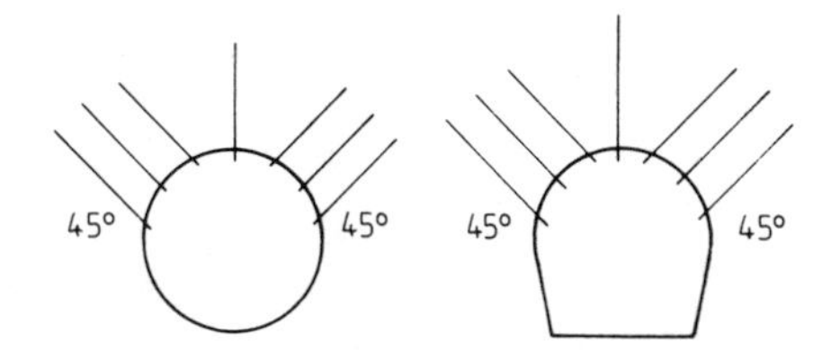

Fig. 10.11 Rock bolting to support tunnel roofing and sides

Table 10.2 Rock bolting for temporary tunnel support (Szechy)

Type of rock	Max. unbridged span (m)	Max. unbridged time	Rock bolt centres spacing (m)	Comments
Firm	–	Unspecifiable	Not needed	No support necessary
Loosening over time	3 - 4	Up to ½ year	2	Rock bolting followed by wire mesh to support falling fragments
Slightly friable	2.5 - 3.5	1 week	1.5	Rock bolting followed by wire mesh or gunited concrete
Friable	1 - 1.5	½ day	1	Rock bolting followed by mesh reinforced gunite up to 30 mm thick
Very friable	0.5 - 0.8	½ hour	0.5	Temporary props followed by rock bolting and reinforced gunited concrete
Immediate light ground pressure	Less than 0.5	1 - 2 min	Unsuitable	Rock bolting not appropriate
Immediate heavy ground pressure	0.1 - 0.2	A few seconds	Unsuitable	Rock bolting not appropriate

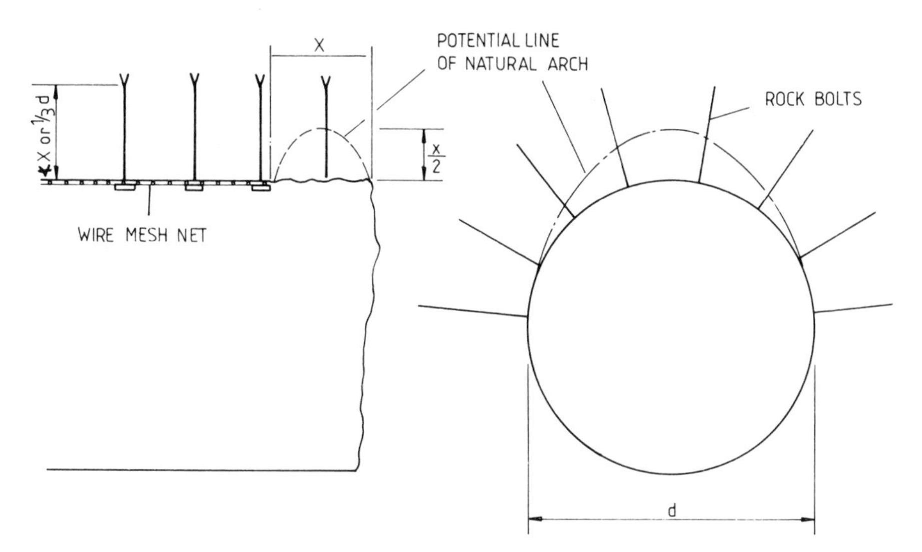

Fig. 10.12 Principles of rock bolting method

If the tunnel heading is advanced a distance x as illustrated in Fig. 10.12 then unless the roof is supported, the fracture zone of rock will begin to fall away and would ultimately settle in a natural parabolic arch, extending about $\frac{x}{2}$ beyond the perimeter of the circular tunnel. The rock bolts should therefore be longer than $\frac{x}{2}$ and generally not less than $\frac{1}{3}$ the tunnel diameter or width (d). The number of bolts and the spacing required to support a given area of tunnel roof and any compressive forces will depend upon the diameter, material and type of anchorage of the bolt.

Types of anchor

Several proprietary systems are available involving a split or spreading device which presses against the walls of the drill hole and grips more tightly as the nut is turned (Fig. 10.13). Others rely on a chemical reaction between resin and quartz-sand. A capsule containing the two ingredients is introduced into the drill hole and mixed by the penetration and turning of the rock bolt (Fig. 10.14).

These methods use relatively small diameter bolts (15 - 30 mm). For more substantial rock bolting with post tensioning, a grouting technique is required (see Chapter 7).

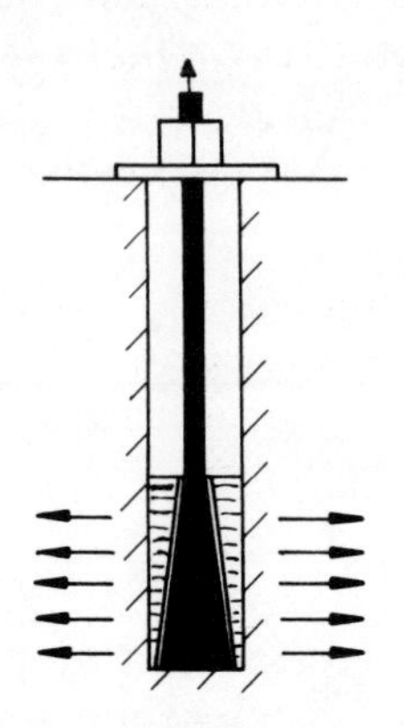

Fig. 10.13 Expanded-base rock bolt

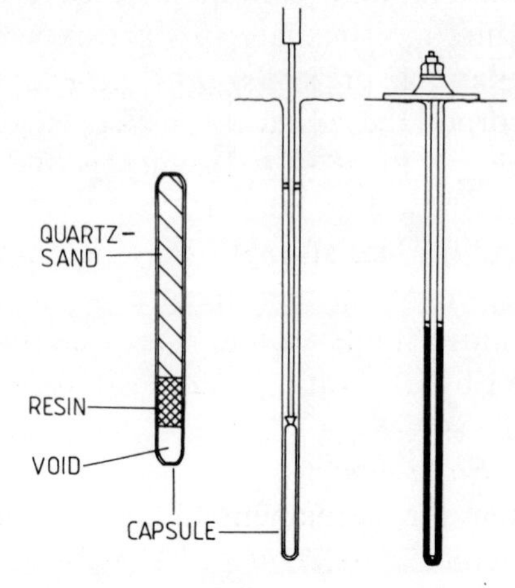

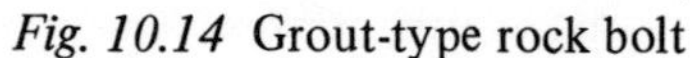

Fig. 10.14 Grout-type rock bolt

Table 10.3 Steel ribs and liner plates for temporary tunnel support (Szechy)

Type of rock	Max. unbridged span (m)	Max. unbridged time	Comments
Friable	1 - 1.5	½ day	Rib support after full face excavation
Very friable	0.5 - 0.8	½ hour	Rib support and a ring of liner plates assembled in segments with subsequent grouting behind the liner plates
Immediate ground pressure	0.1 - 0.2	Less than 2 mins.	Rib support and a ring of liner plates assembled in segments, followed by an overlay of reinforced gunite

(c) *Stiffening ribs and liner plates*

Table 10.3 illustrates that for very friable or shattered rock and where immediate heavy ground pressure is to be encountered, neither guniting nor rock bolting is a suitable means of providing temporary support and in such conditions propping with steel ribs is required. When the rock will stand unsupported long enough to install a complete unit or ring, then full face excavation is possible. Otherwise the face should be excavated in segments and parts of the rib temporarily propped until a complete unit can be formed (Fig. 10.15). In friable rocks which require intermediate support between ribs or where immediate ground pressure is present during excavation, liner plates (Fig. 10.16) may be incorporated as shown in Fig. 10.17. The voids between the plates and rock face may subsequently be grouted to distribute the loads uniformly (Fig. 10.18).

Additional strength may be obtained by overlaying the liner plates, ribs, etc. with reinforced gunited concrete as illustrated in Fig. 10.19. In more stable ground it is often possible to use the liner plates without the ribs, excavating a full face to the width of a liner plate.

(d) *Full lining*

When the permanent lining is designed in either cast iron or precast concrete units, as is occasionally the case, it can then perform a similar function to liner plates. The usual method is to build a ring of linings

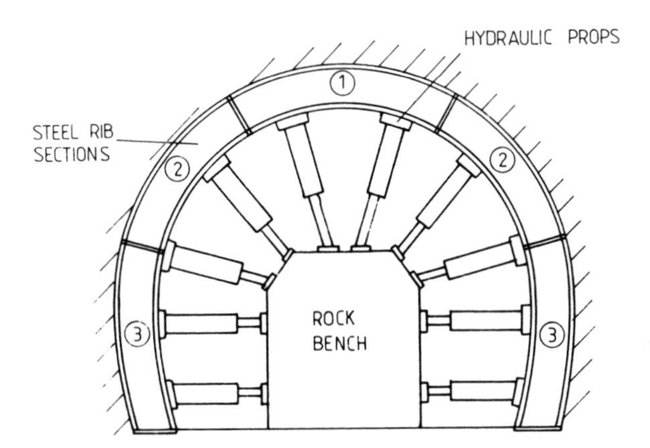

Fig. 10.15 Propping very unstable rock with props and ribs

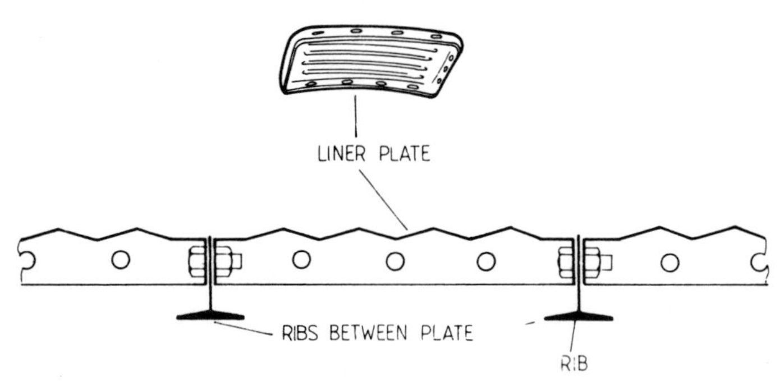

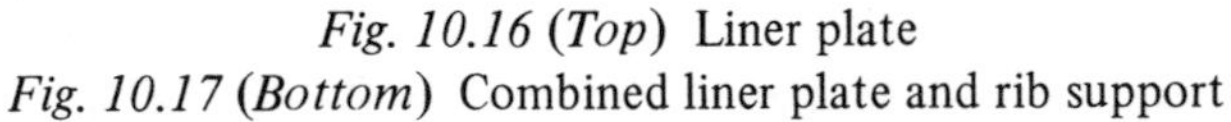

Fig. 10.16 (*Top*) Liner plate
Fig. 10.17 (*Bottom*) Combined liner plate and rib support

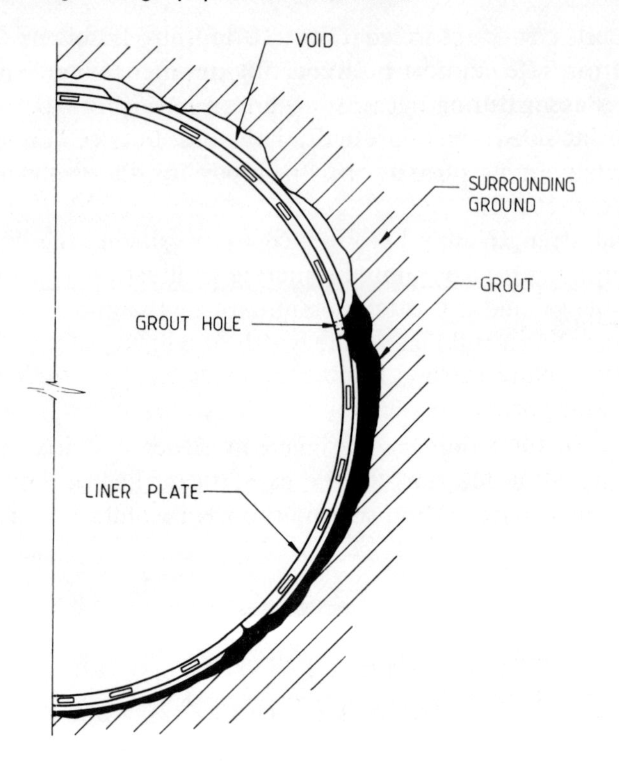

Fig. 10.18 Grouting the void behind the lining

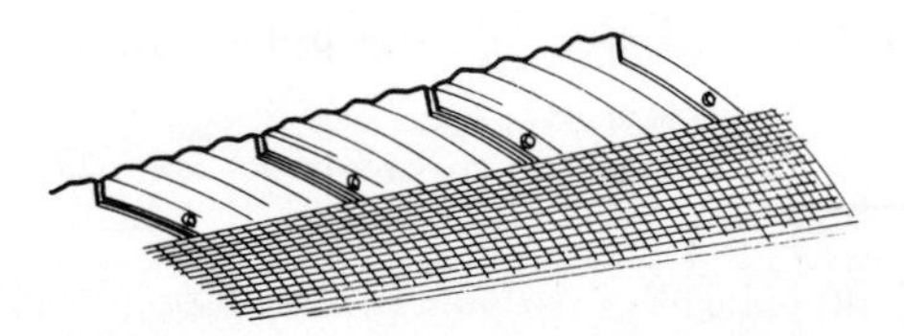

Fig. 10.19 Reinforcing the ribbing with reinforced sprayed concrete

from the bottom upwards. It is therefore necessary to adopt a full face method of excavation. If heavy, the lining units can be positioned with special equipment (see Fig. 10.37).

The method may be used in conjunction with rock bolting to increase working space so that lining and excavation areas do not become too congested.

Tunnelling in soft to medium ground

Liner plate method

In relatively stable ground, where large inflows of water do not disturb the soil particles, full face excavation to a depth equal to the width of a lining panel can be carried out, followed by immediate extension of the lining ring as described earlier under the Full Lining method. Stiffening ribs may be included where ground conditions dictate. Unfortunately, this temporary lining system, which is very speedy, must remain in place whilst the permanent lining is erected, and is therefore unrecoverable.

Shield tunnelling

Where the liner plate method is uneconomical or where the ground will not stand unsupported long enough to enable the permanent lining to be erected, then the shield technique is likely to be selected. The shield acts as a steel casing tube, which is pressed into the heading in front of the lining, thus providing protection whilst the face is excavated. This procedure is shown in Fig. 10.20. In order to reduce resistance to

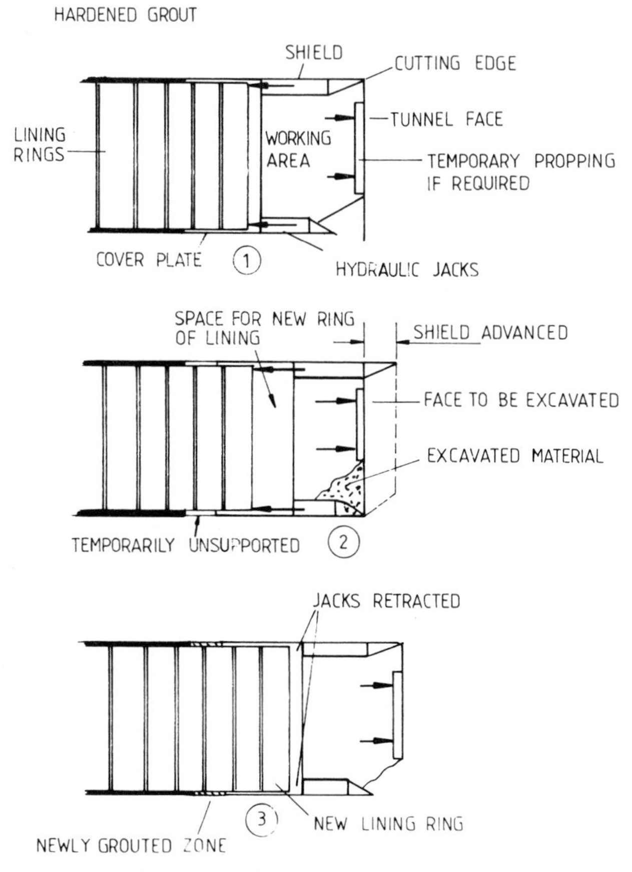

Fig. 10.20 Shield tunnelling method

penetration, as much of the face as practicable is excavated in front of the cutting edge. Hydraulic jacks at the rear use the lining as the source of reaction to push the shield forward a distance equal to the width of a ring of lining. The lining is subsequently extended under the protection of the shield and the space left by the thickness of the shield plating is finally filled by grout. (Good practice usually requires grouting after the erection of each ring, to reduce the possibility of soil movement).

The excavation of the face is usually carried out manually with power-assisted hand tools (Fig. 10.21) or, where the expense can be justified, with mechanical excavating equipment such as a backhoe (Fig. 10.22).

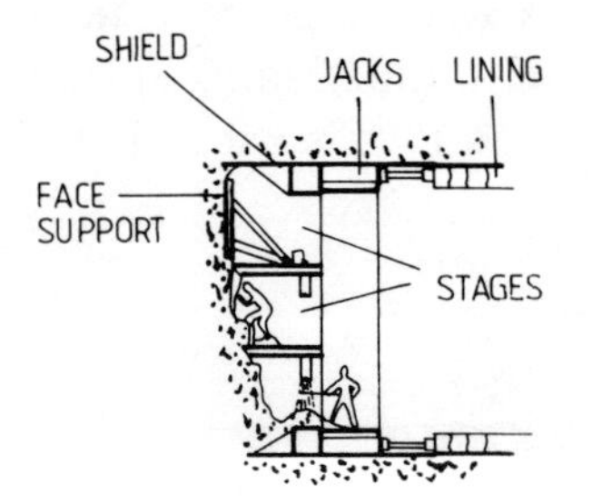

Fig. 10.21 Manual excavating method illustrating a shield with working platforms

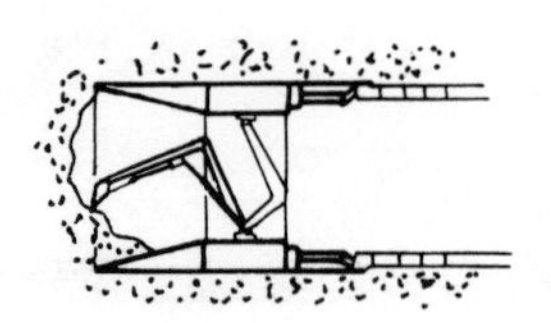

Fig. 10.22 Excavating with mechanised equipment using the shield method

Where blasting is required, for example in hard rock, the shield method is usually avoided because of the technical problems associated with steering. The tendency would be for the shield to wobble in a fully excavated face. The shield method is suitable for tunnel diameters of 4 - 12 m. For the larger tunnels in this range staging is necessary to carry out the excavation work (see Fig. 10.21) and to provide foundation bases to prop the working face. Indeed, in a freely-flowing soil a special bulkhead shield is often required (Fig. 10.23).

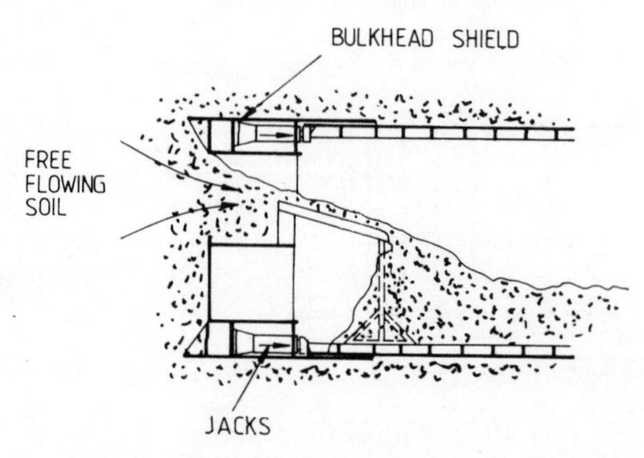

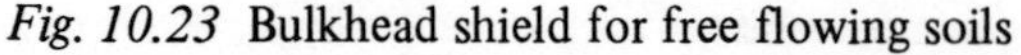

Fig. 10.23 Bulkhead shield for free flowing soils

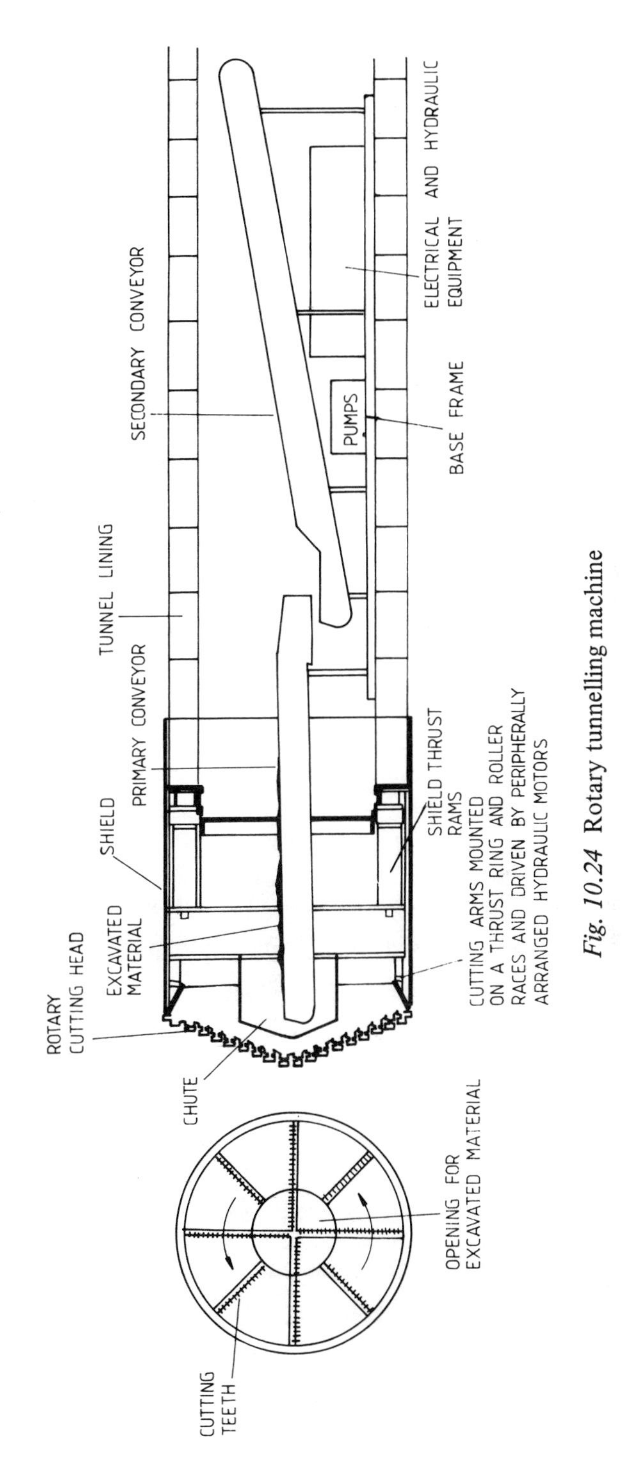

Fig. 10.24 Rotary tunnelling machine

Thrusting forces required in the shield method

For a shield without stage levels, a thrust against the lining of up to 100 tonnes/m of shield circumference may be necessary. Where obstructions are encountered, considerably higher force could be needed. When the shield is fitted with staging levels, the required thrust is likely to be 300 – 400% more than the single unstaged shield.

Rotary tunnelling machines

Rock cutting machines (*Fig. 10.24*)

The demand for more economical tunnelling methods has gradually led to the development of rotary machines. Where uniform soil or rock is present and the length of the tunnel exceeds about 2 km, the high capital cost of the equipment can be justified by the increased production performance. However, if the nature of the ground is likely to change frequently, especially in soft ground where boulders, flowing sands, gravels, etc. might render the cutting head unsuitable for such a variety of material (the rotary machine requires a free standing face to attack), then a more traditional shield method would probably be chosen.

The rotary tunnelling machine combines a conventional shield with a rotary cutting head. The cutting head consists of 3 to 8 radial arms fitted with chisels or discs and is rotated at 5 – 8 cycles per minute. The arms are mounted on a thrust ring and roller races and are driven by compact hydraulic motors arranged around the periphery. The force at the cutting edge is provided by the reaction of thrust rams against the lining. Excavated material is directed centrally through the cutting head and transported by conveyor to some convenient point away from the working area. A complete cycle of operations is shown in Fig. 10.25.

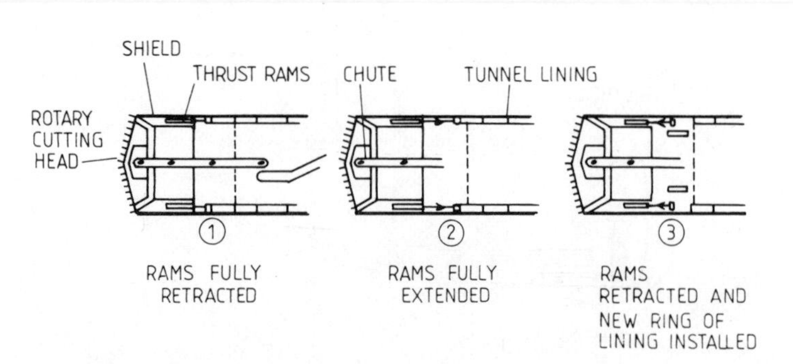

Fig. 10.25 Method of working with the rotary tunnelling machine in rock using a tail shield

In hard and more stable rock the tail shield is not required. Also, the lining may be of much lighter construction and could not withstand the high jacking forces. The reaction must therefore be taken against the tunnel sides as indicated in the sequence shown in Fig. 10.26. Diameters ranging from 2 - 10 m are available. Thrusts of up to 2 kN/mm of cutting head diameter can be delivered for cutting through hard rock.

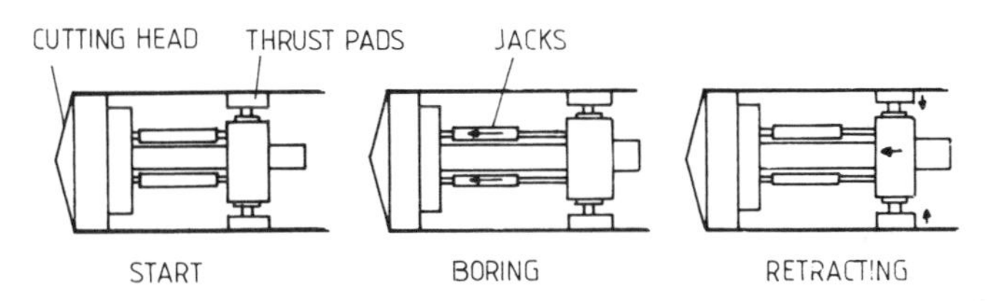

Fig. 10.26 Method of working with the rotary tunnelling machine in stable rock

Soft ground machines

Tunnelling in granular soils, such as water-bearing sands and gravels, usually presents severe difficulties to conventional shield-drive methods. For example the ground would have to be extensively grouted, de-watered, or the work carried out in compressed-air. Also, the face must be propped to avoid collapse, thereby causing difficulties in excavating and in pushing the shield forward. However, the alternative of using a rotary shield requires a face that will stand vertically to allow the cutting action and spoil removal to take place efficiently. This problem led to the proposal of using bentonite slurry to provide the support and subsequently a machine was manufactured by R.L. Priestley Ltd. (Fig. 10.27).

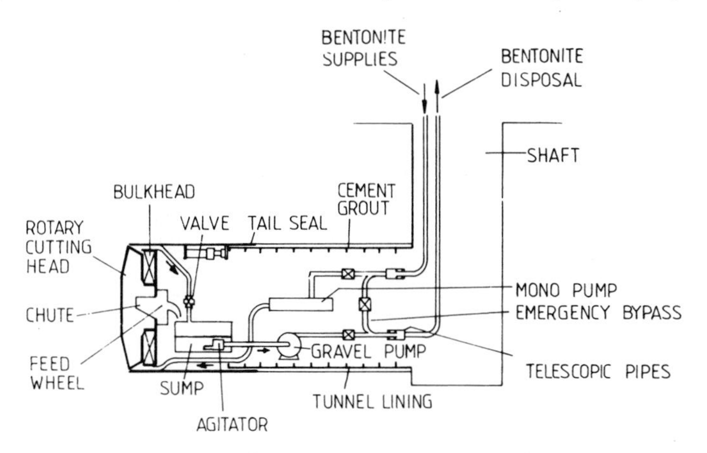

Fig. 10.27 Soft-ground rotary tunnelling machine

It is similar to a soft rock rotary-cutting machine, but the cutting head operates against the face in a chamber of circulated bentonite slurry. The mixture of spoil and slurry is continuously discharged through the centre of the cutting head into a sump and pumped away for disposal.

A gang of 5 - 6 workmen is required to operate the machine, fix the lining, handle the spoil, etc.

Selection of the cutting head

The methods available to deal with different rock hardnesses are shown in Table 10.4.

Table 10.4 Cutting heads for tunnelling machines

	Soft	Medium	Hard	Very hard
	Shale Clay Limestone Granular soils	Limestone Sandstone Sandy shales Porphyries Iron Ore	Silicon Limestone Dolomite Granite Iron Ore Quartzite	Iron formation Quartzite Hard igneous rocks
Compressive strength (N/mm^2)	up to 50	100	200	300
Cutting heads	Teeth or picks	Roller discs	Roller buttons	Blasting with explosives

(i) *Teeth or picks* (*Fig. 10.28*)

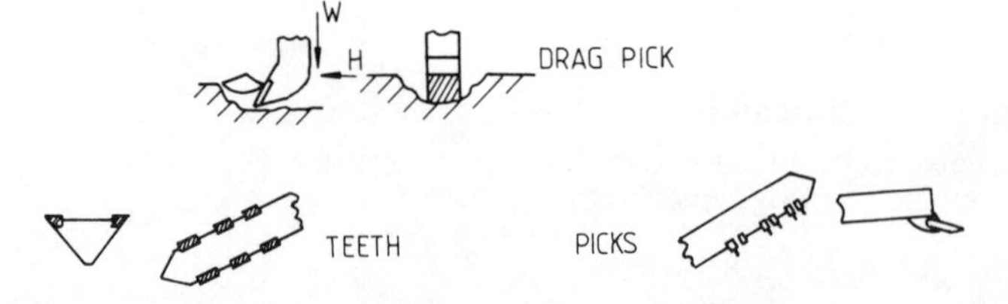

Fig. 10.28 Teeth and picks for soft rock

Tungsten carbide tipped teeth are generally mounted on radial arms to form the cutting head which is either rotated or oscillated into the heading.

(ii) *Roller discs* (*Fig. 10.29(a)*)

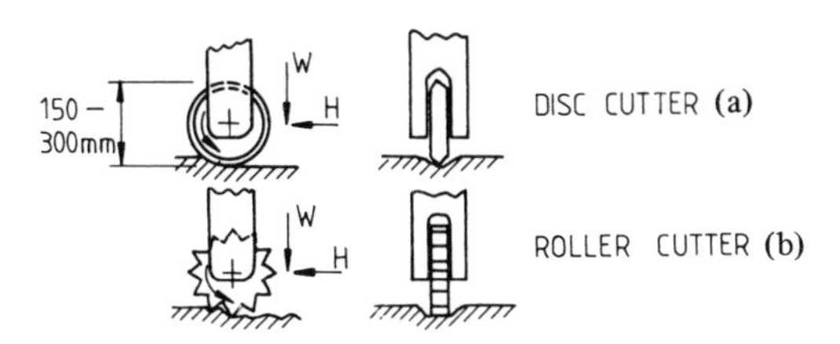

Fig. 10.29 Discs and cutters for medium rock

These are arranged symmetrically on a strong base and a heavy force is applied to the cutting head, causing the rock to shatter and splinter. The alternative is roller cutters (Fig. 10.29(b)), which produce slightly finer rock fragments.

(iii) *Roller buttons* (*Fig. 10.30*)

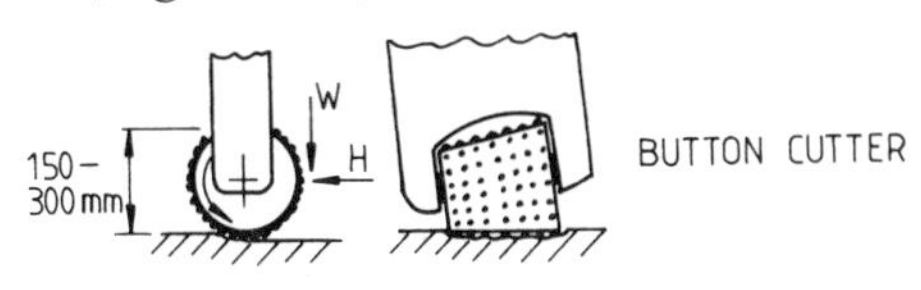

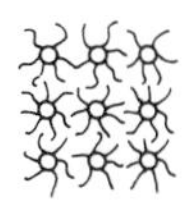

Fig. 10.30 Roller buttons for hard rock

Hardened metal buttons mounted on a conical roller are pressed with great force against the heading and rotated, causing fracturing of the rock surface.

(iv) *Blasting*

At present, mechanical techniques are inadequate for excavating very hard rock and explosives must be used to loosen the material. It is then loaded with conventional machinery (Fig. 10.1). The use of explosives is discussed in Chapter 4.

Boom cutter/loader for rock tunnelling

As an alternative to rotary machines or blasting, the boom cutter/loader can be used in rocks and hard soils with compressive strengths

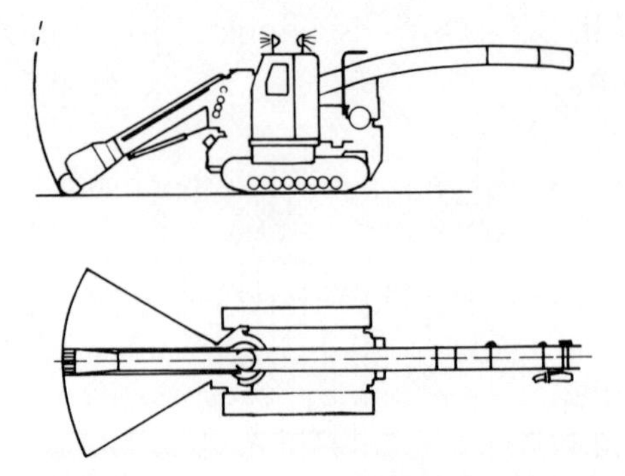

Fig. 10.31 Boom cutter/loader machine

of up to 150 N/mm², i.e. medium/hard. However, drilling and blasting is likely to be more economical for rocks stronger than 50 N/mm². The machine (Fig. 10.31), about 100 kW power, can be mounted on rails or tracks and works against the face with a contact pressure of about 0.1 N/mm² using a series of rotating teeth or blades. The material is immediately transferred by conveyor to a convenient loading point. About 18 m³/h of soft rock can be removed.

Production performance

Rock tunnelling

(i) Explosives and modern propping systems (see Tables 10.1 – 10.3) – 1 – 7 m per 8 h shift.
(ii) Classical methods – 0.5 – 2 m per 8 h shift.
(iii) Tunnelling machines – Up to 15 m per 8 h shift, depending upon rock hardness and rate of lining erection.

Soft ground tunnelling

(i) Tunnelling machines
  (a) Up to 20 m per 8 h shift in uniform soil such as clay and very soft rock.
  (b) 2 – 5 m per 8 h shift in water-bearing granular soils.

(ii) Shields
  (a) hand digging
    (i) Uniform soil such as clay – 4 – 8 m per 8 h shift
    (ii) Poor ground – 1 – 4 m per 8 h shift

(b) mechanical excavation, e.g. backhoe	Outputs similar to tunnelling machines.

Notes

(i) The number of men needed for excavating, and/or the machinery required to load material depend upon the working area available, e.g. 2 m diameter heading can accommodate two face miners.
(ii) In soft ground rotary machine tunnelling the rate of lining erection is often the limiting factor, e.g. in non-bolted segments, about 3 m/h of advance is likely to be the maximum rate.
(iii) It should also be noted that the wide range of output rates in rock tunnelling is dependent upon the amount of propping required, incoming water, etc.
(iv) The shield can take from 1 - 4 weeks to assemble and position in the heading, depending upon its size and associated mechanical equipment.

Lining erection

A tunnel usually requires permanent lining which, wherever possible, is utilised for temporary support during advancement of the heading. In good practice, the lining is followed up a short distance from the excavation face. There are three common forms of lining used in tunnel design:

(i) Segmental
(ii) In situ reinforced concrete
(iii) Masonry

(i) Segmental

(a) Cast iron units (Fig. 10.32) approximately 1 - 2 m long × 0.5 - 1 m wide when bolted together provide a strong lining which is able to resist both the heavy jacking forces from the shield and external loads imposed by the surrounding ground. By making the joints between butting flanges of lead wire or similar and using bituminous washers to cover bolt holes, the lining can be reasonably sealed against water penetration.
(b) Precast concrete segments (Fig. 10.33) have been introduced to avoid the high cost of using cast iron. Joints are made tongue and groove and caulked with rubberised bituminous strips. The

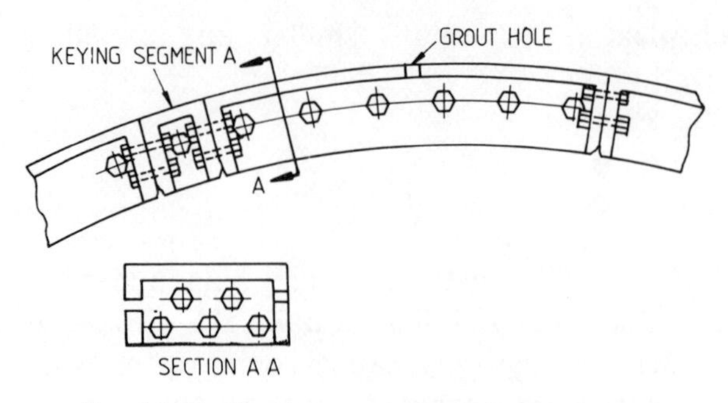

Fig. 10.32 Cast iron permanent lining segment

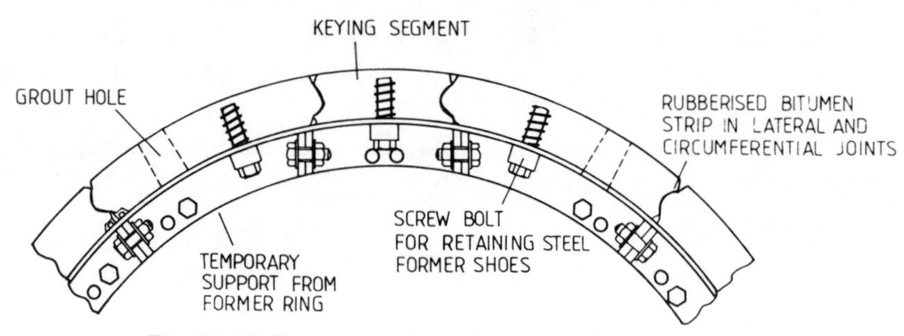

Fig. 10.33 Precast concrete permanent lining segment

segments may be bolted together in a similar way to cast iron units, but where uniform loads can be expected, bolting can be eliminated, thereby providing some flexibility for adjustment to ground movements. However, as a result, the units must be temporarily supported until the keying section at the top can be placed in position (Fig. 10.34). This problem can be avoided by

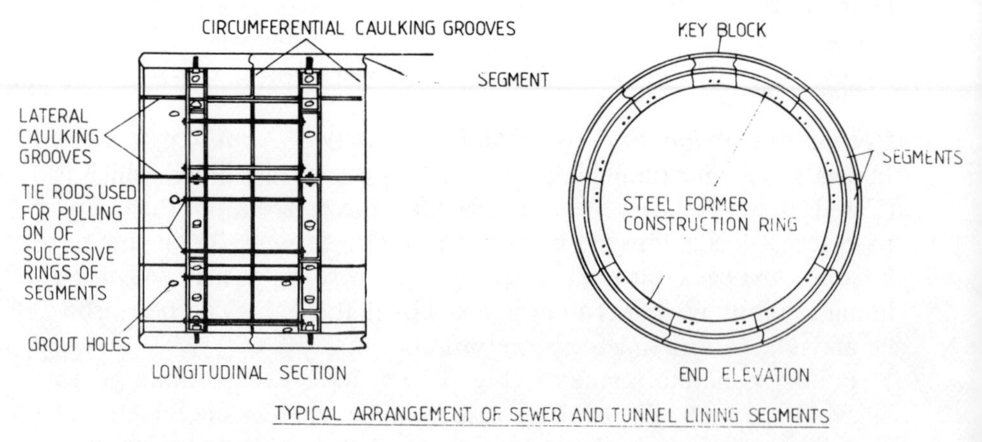

Fig. 10.34 Temporary support for non-bolted precast concrete lining

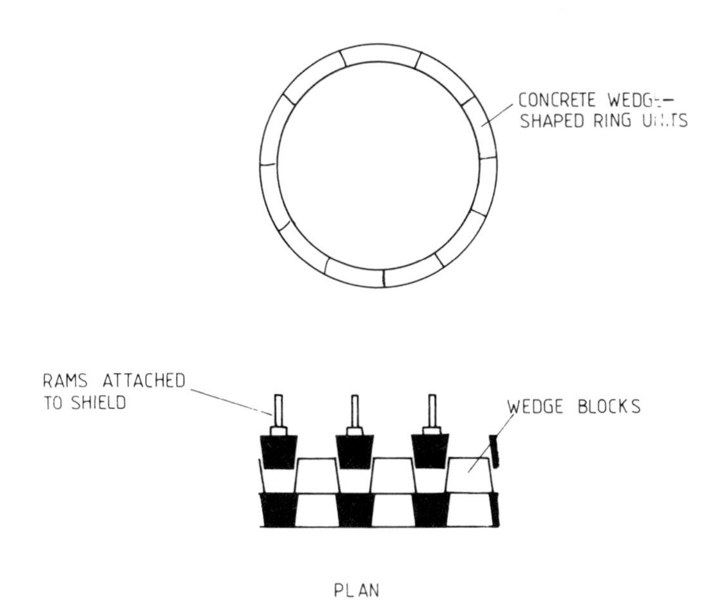

Fig. 10.35 Wedge block system using precast concrete lining segment

using a wedge block system shown in Fig. 10.35. Alternate blocks are positioned and are individually held in place by arms on the shield or rotary machine. The remaining blocks are subsequently pushed between the taper spaces, again by rams, to produce a tightly fitting ring. No bolting is required and installation rates of up to 3 m/h can be obtained with mechanical handling, and it is therefore a favoured method when using rotary tunnelling machines, where high rates of progress are needed to match the rate of advance of the machine.

Other materials have been tried, including steel; in particular, steel liner plates (see Fig. 10.16) provide an economic alternative where a heavy lining is not required, such as in rock tunnelling, or they can be incorporated into an in situ concrete permanent lining.

Method of erection

In small diameter tunnels up to 2 – 2.5 m, segments can be manhandled (Fig. 10.36), but larger tunnels require special machinery (Fig. 10.37) to place the individual units. It is normal practice to work from the bottom, finishing with the key section at the top. The void between lining and tunnel wall is then grouted so that the ground pressures can be taken up by the ring of segments.

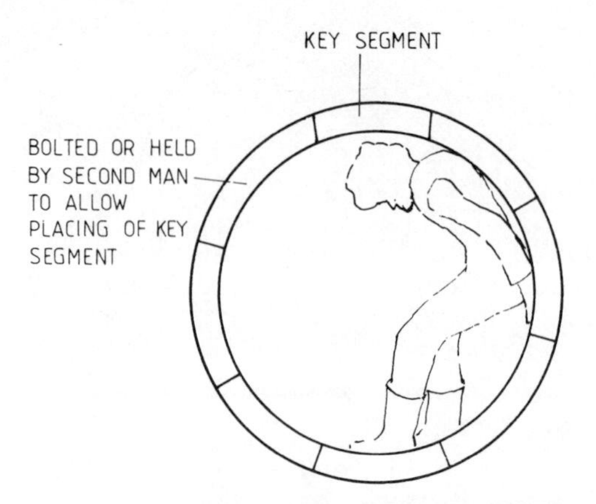

Fig. 10.36 Manual method of positioning lining segments

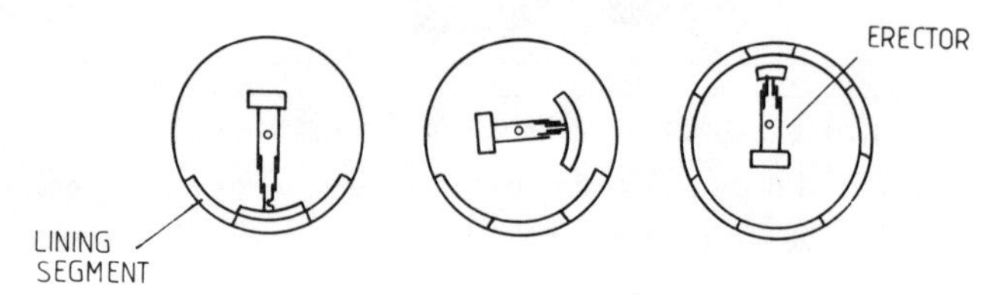

Fig. 10.37 Mechanised method of positioning lining segments

(ii) In situ reinforced concrete

Frequently in rock tunnelling the roof is able to stay unsupported, or at least is stabilised with rock bolts or ribbing. For such tunnels an expensive cast iron lining is not required, as reinforced sprayed concrete may be adequate. If the lining requires greater strength or a preformed shape, then normal reinforced concrete can be cast behind specially designed travelling formwork and falsework (Fig. 10.38).

(iii) Masonry

In earlier times, when using classical tunnelling methods, the complete lining was produced in masonry brickwork, which required temporary falsework. Today, such a lining is mostly used to provide a protective coating of other lining systems for tunnels exposed to corrosive elements.

Stabilisation methods

So far, problems caused by the ingress of water into the heading, or the changes encountered in soil texture and stability have been avoided.

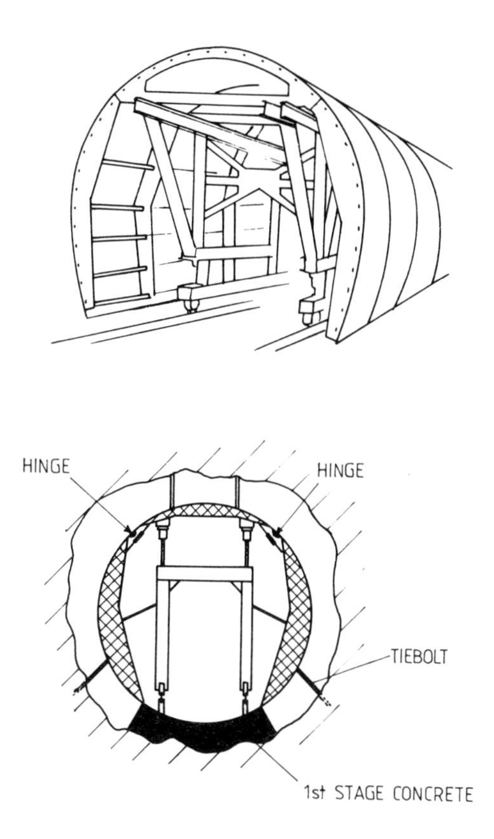

Fig. 10.38 Travelling formwork for in situ concrete lining

Such difficulties can generally be contained by the use of grouting or freezing techniques, as described in Chapters 7 and 9. A small diameter pilot tunnel driven prior to the main heading often enables lenses of water-bearing soil, rock fissures, etc. to be filled before being exposed by the full heading. However, where incoming water is likely to be a problem through much of the drive, the grouting may be too expensive and the use of compressed-air should then be considered. In general, inflows of water in excess of 2 - 3 lps per m^2 of tunnel face would require containment.

Compressed-air for tunnels (including caissons and diving work)

The pressure of water at the tunnel level can be counter-balanced by pressurising the working area, as shown in Fig. 10.39. However, in order to have sufficient head at the sole of the tunnel, excess pressure exists at the roof and some leakage is inevitable (Fig. 10.40), especially in granular soils. If the leakage is excessive, a 'blowout' may occur, causing

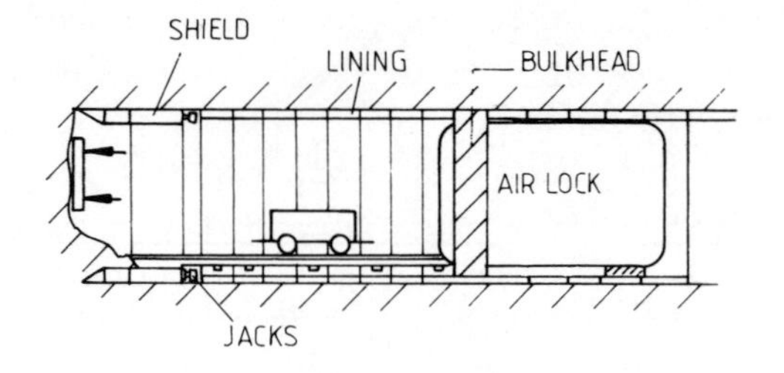

Fig. 10.39 Working in compressed-air

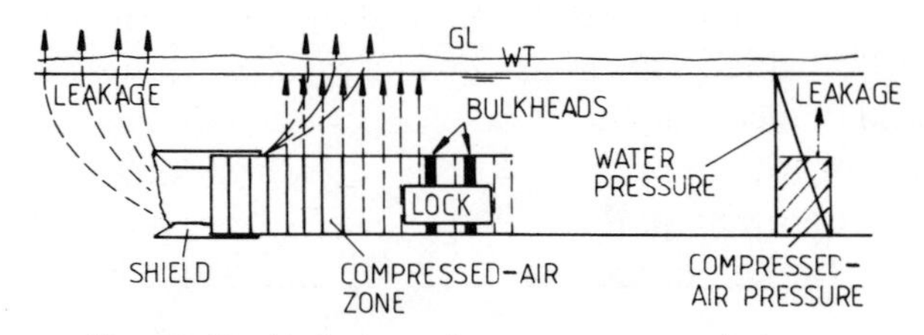

Fig. 10.40 Air leakage from a compressed-air zone

a sudden rush of water into the heading. This possibility can be reduced by covering the ground surface above the line of the tunnel with a layer of clay (Fig. 10.41). In more severe cases extensive grouting may be the only alternative and in any case ample spare compressed-air supply should always be available. Hewett-Johanneson recommends provision of about 5 m^3 free air per min per m^2 of face area for cohesive soils and 10 m^3 per min per m^2 in granular soil. The compressed-air in the working zone is built up against a bulkhead, through which both men and materials need access to the working area by means of an air lock. This consists of a small chamber, called a man-lock or decompression chamber (see Fig. 10.39) within which the air pressure can be reduced to or raised from atmospheric, depending upon whether personnel are

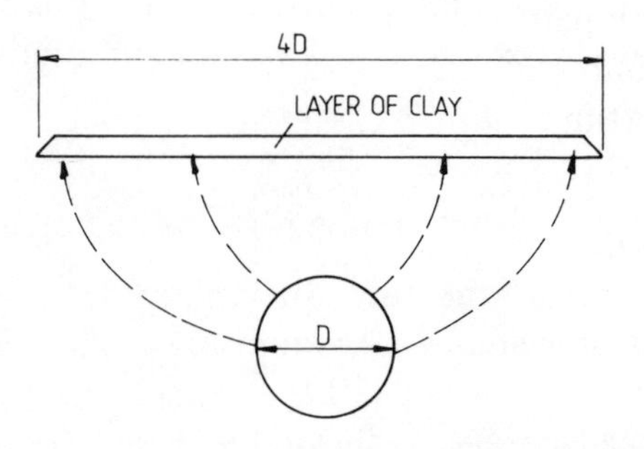

Fig. 10.41 Method of reducing loss of compressed-air

entering or leaving. The procedures are known as compressing and decompressing, and must be carried out in strict sequence, otherwise serious injury to the miners will result.

In tunnelling schemes using compressed-air, it is good practice to have two adjacent air-locks, one for personnel and one for materials. The latter does not require the elaborate decompressing process. However, space is often at a premium and only a single lock may be possible. In compressed-air working, pressures up to 50 p.s.i. (3.5 bar) above atmospheric have been used, but a more typical maximum is about 35 p.s.i. (2.5 bar). The compressing and decompressing procedures are laid down in 'The Work in Compressed-Air Special Regulations' (1958), which specifies decompressing times, lengths of working periods, medical checks and other requirements. However, medical criticism of these regulations has prompted the Medical Research Council to recommend new tables, known as the 'Blackpool Decompression Tables', after the first contract for which they were adopted.

Permission to use these new tables should be sought from the Chief Inspector of Factories.

Because of the rigorous procedures necessary for working in compressed-air, the hours a workman can actually spend at the work face are fewer than those of a normal shift, although pay must be the same as for a normal length shift. Consequently this method is extremely expensive.

Dangers of working in compressed-air

The most visible dangers of working in compressed-air are generally associated with inadequate decompression and fall into two categories – acute and chronic decompression sickness.

Acute or immediate decompression sickness

Type 1 – mild limb pain (the 'niggles')
severe limb pain (the 'bends')
skin mottling and irritation (the 'itches')

Type 2 – vomiting, with or without stomach pains
vertigo (the 'staggers')
tingling and numbness of the limbs
paralysis or weakness of limbs
choking
severe headache
visual defects

angina, irregular pulse
collapse, hypertension
lung cysts
coma
death

Chronic or long-term decompression sickness

(i) Nervous and psychiatric forms, including paralysis
(ii) Disease to the bone joints

Medical evidence indicates that decompression sickness is primarily caused by nitrogen bubbles being trapped in the blood and tissues, following rapid decompression and that very slow decompression prevents this happening. However, an understanding of the medical problems is not complete at present and research is continuing.

Tunnel ventilation

Air must be supplied to both workmen and machinery operating in the tunnel, so that the level of oxygen content does not fall below about 20%. A means of extracting carbon dioxide produced from exhaling, explosives, internal combustion in engines, etc. must also be provided.

The following values are recommended:

Workmen - 2 - 5 m^3 per min of fresh air per worker.
Diesel engines - 2 - 3 m^3 per min per kW of power generated.
Explosives - Several formulae have been developed relating to the weight of explosives used, gases produced, minimum waiting time required after the explosion before the next explosion, number of workmen involved, etc., but a practical guide is:
$Q = 10 \times$ cross-sectional area of tunnel in m^2
Q is in m^3 per min of fresh air.

Methods of ventilation

Fresh air can be introduced to the work face by blowing the air along a pipeline of 300 - 500 mm diameter with its intake near to the tunnel entrance (Fig. 10.42). The displaced used air simply percolates back along the tunnel and is expelled at the entrance. Alternatively, in the reverse of this system, contaminated air is drawn away from the work face and fresh air pumped in at the entrance to the tunnel (Fig. 10.43). In very long tunnels intermediate shafts designed to provide permanent

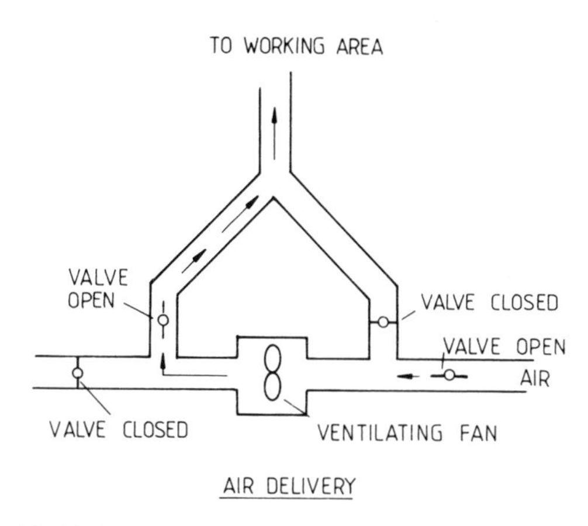

Fig. 10.42 Ventilation system using air delivery method

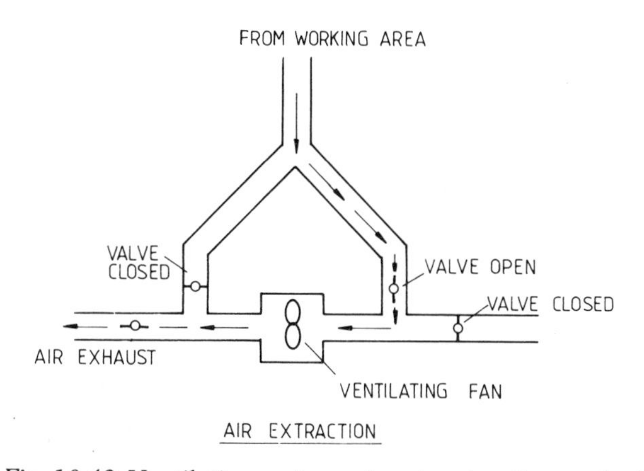

Fig. 10.43 Ventilation system using air extraction method

ventilation can be used to reduce the length of pipe runs and so limit the pressure drop caused by friction. A typical ventilation fan is shown in Fig. 10.44 with its corresponding characteristics in Fig. 10.45.

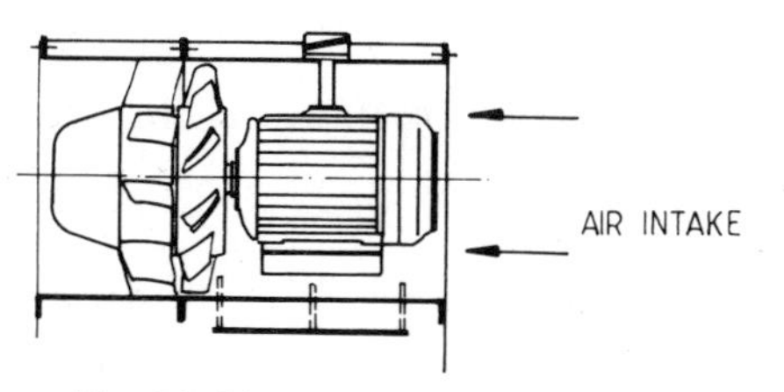

Fig. 10.44 Typical ventilation fan

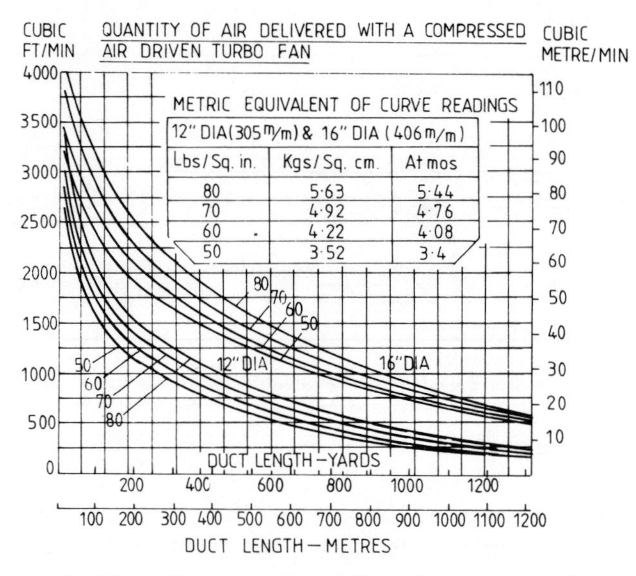

Lbs/Sq. in.	Kgs/Sq. cm.	Atmos
80	5·63	5·44
70	4·92	4·76
60	4·22	4·08
50	3·52	3·4

Fig. 10.45 Quantity of air delivered with a turbo fan

Clearly, as the length of pipe run is increased, the power size of the blower must also be stepped up to achieve the desired delivery or extraction rate.

Other developments in tunnelling

(i) Pipe jacking

There exists a demand for short lengths of small diameter tunnel for use in sewers, tubes, subways, ducts, sleeves, etc. Typically, for distances up to about 100 m and diameters in the range 1 – 4 m, the pipe jacking method is proving economical in soft/medium hard ground. The technique uses either concrete or steel pipes.

Construction principle

The principle of pipe jacking requires 3 – 5 m sections of pipe to be pushed from a jacking pit. Excavation takes place as driving proceeds and may be done by hand or machine (Fig. 10.46). The first section of pipe is fitted with a cutting shield and minor directional control is achieved with secondary rams. For long drives e.g. 500 m in clay, intermediate rams can be introduced to limit the frictional resistance within smaller sections, and a film of bentonite lubricant may be injected between the pipe and soil to reduce the frictional forces

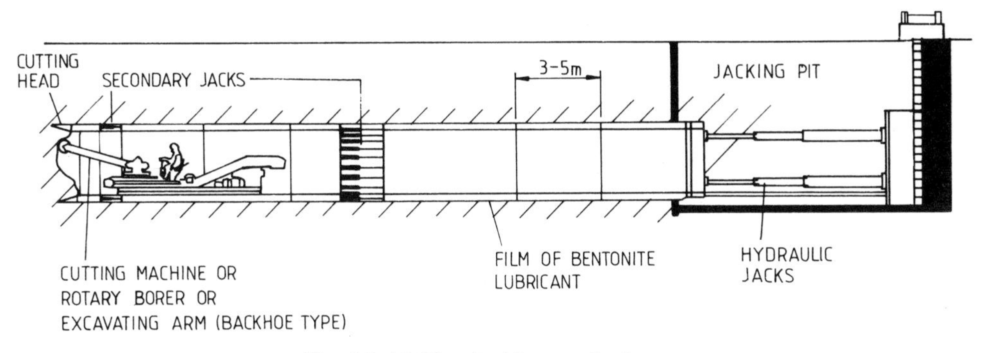

Fig. 10.46 Pipe jacking method

further. Where a jacking pit cannot be provided, for example when driving through embankments, a slightly different procedure is necessary. Here, a series of holes is first bored, passing through the full width of the embankment. Tie bars are then fed through the holes from the jacking bulkhead to a temporary reaction face (Fig. 10.47). Jacking can then proceed as described.

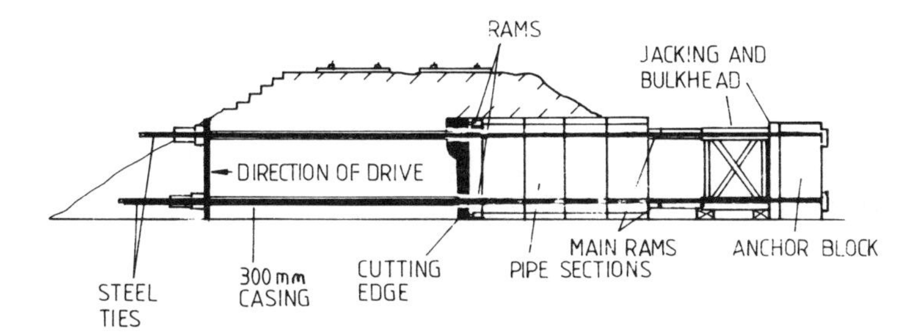

Fig. 10.47 Pipe jacking using a tension held bulkhead

Details

Accuracy	50 mm in 100 m
Production (approx.)	1 – 3 m advance per 8 h shift (2 miners, 2 labourers and a crane driver)
Diameter	1 – 4 m
Drive length	Currently about 600 m max.
Main rams	3000 – 3500 kN thrust each
Secondary rams	500 – 800 kN thrust each

Note: (i) Up to 20 m per hour shift has been achieved with mechanically-operated backhoe-type shields in dry uniform soils, e.g. London clay.
(ii) Water-bearing soils and flowing soils may cause serious production difficulties.

Auger boring

The pipe jacking technique is unsuitable for the installation of pipes and ducts of less than approximately 1 m diameter, because of the difficulties of carrying out the excavation, and in this case an augering method is required (Fig. 10.48).

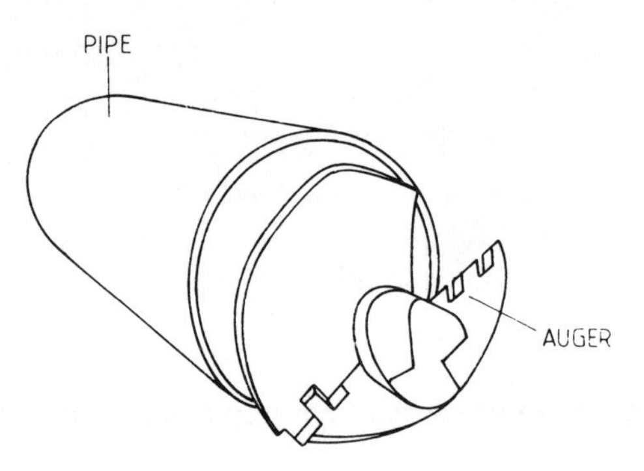

Fig. 10.48 Rotary boring method

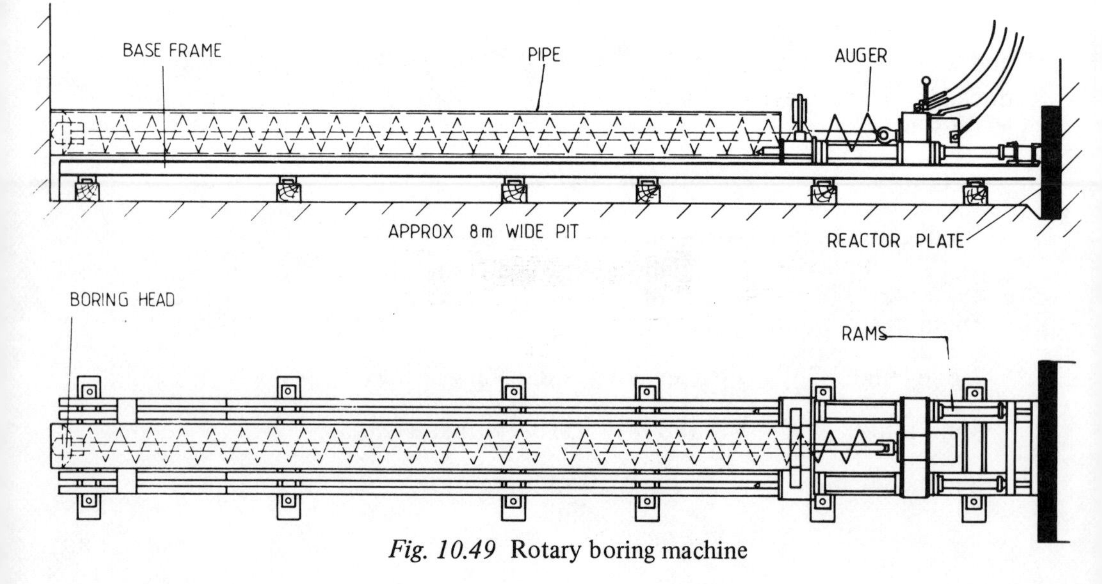

Fig. 10.49 Rotary boring machine

The equipment consists of a base frame, auger and rams (Fig. 10.49). Either steel sleeves or concrete pipes are thrust through the ground as the heading progresses forward. The technique is essentially used in soft ground, but it has been used in soft rocks, such as chalk and sandstone. Equipment specifications shown in Table 10.5 indicate the capabilities and approximate boring distances of machines (depending upon soil conditions).

Details

Rotation speed	30 - 60 r.p.m.
Accuracy	1° in 30 m
Diameters	Up to 1200 mm
Drive length	150 m max., limited by thrust needed to overcome friction
Output	Up to 10 m per 8 h shift in soft/medium uniform soils

Note: For both pipe jacking and auger boring, up to 1 – 2 weeks may be required to prepare and position the equipment in the jacking pit.

Percussive boring

A new technique developed for installing pipes of 200 mm diameter and less has been developed by the Grundamat Company. The method requires the pre-forming of a hole slightly larger in diameter than the pipe, through which the pipe is subsequently fed.

The equipment consists of a piston-driven chisel head (Fig. 10.50) operated by compressed-air (minimum 6 bar). The head is lined up and started from a cradle (Fig. 10.51). Compressed-air is then introduced and the device simply punches its way through the ground.

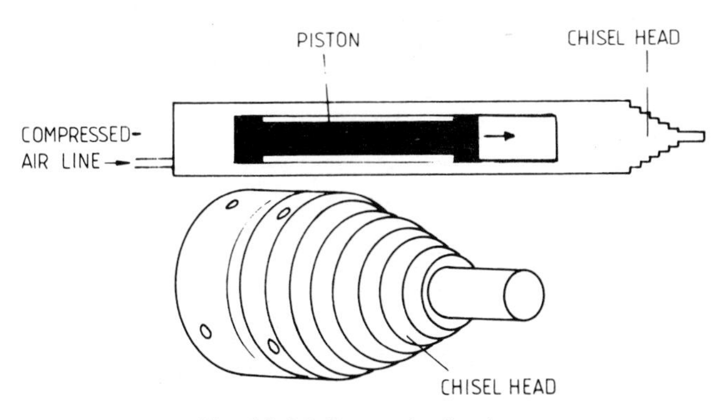

Fig. 10.50 Percussive boring

Table 10.5 Boring machine specifications as distance in metres

	Auger diameter (mm)												
Thrust power	80	150	200	300	400	500	600	750	900	1050	1200	1350	1500
46 kN 5 kW	30	30	27	26	15								
46 kN 7.5 kW	36	30	26	21	15								
130 kN 15 kW	91	84	69	49	34	27	23						
270 kN 15 kW	91	91	91	84	58	44	35						
270 kN 50 kW			91	91	72	61	46	30	23				
540 kN 50 kW			91	91	91	85	67	46	33				
560 kN 60 kW				91	91	80	73	56	44	38	33		
1200 kN 60 kW				106	106	91	91	85	68	57	45		
2700 kN 75 kW						152	152	152	144	114	99	76	61

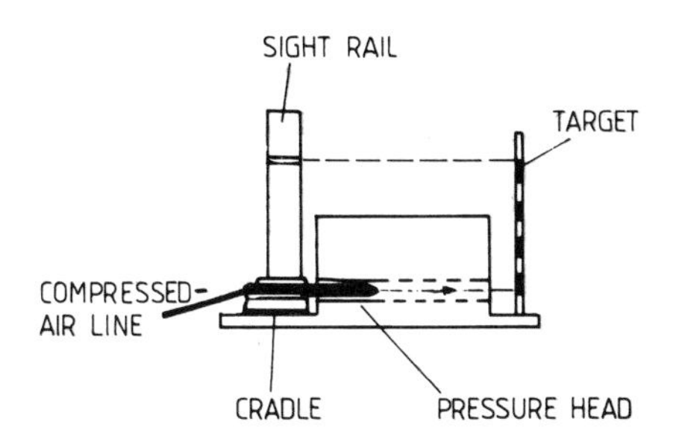

Fig. 10.51 Directing the percussive boring machine

Obstructions such as bricks or stones are crushed or pushed aside (the head can be reversed out if necessary). On completion of the boring, the pipe sections are pushed into position. Some thrust assistance may be necessary for lengths exceeding approx. 15 m. (Fig. 10.52).

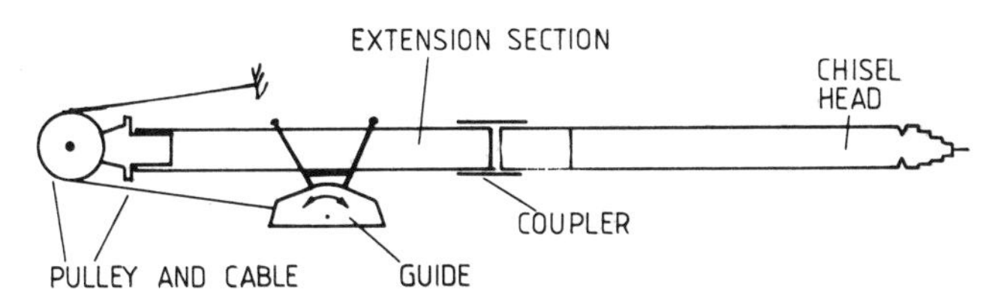

Fig. 10.52 Pulley and tackle to give extra thrust in percussive boring

Immersed tubes

Units up to 100 m X 40 m are laid end to end to form the tunnel (Fig. 10.53). The procedure consists of:

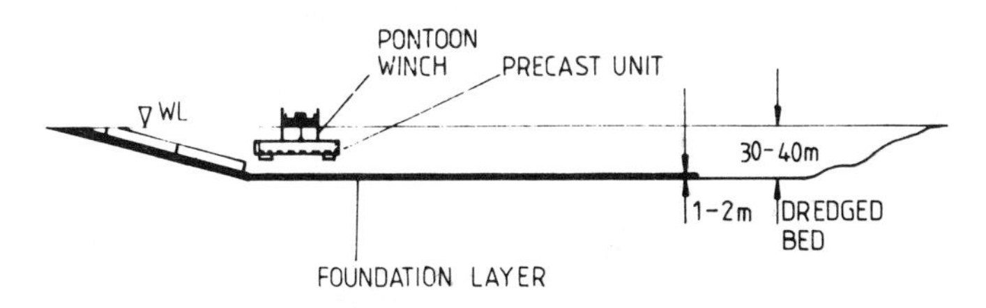

Fig. 10.53 Principle of the immersed tube tunnel method

(i) dredging the trench in which the tunnel is to be placed
(ii) preparing either a permanent screeded bed (Fig. 10.54) or piled foundation (Fig. 10.55), depending upon the soil, to support the tunnel sections

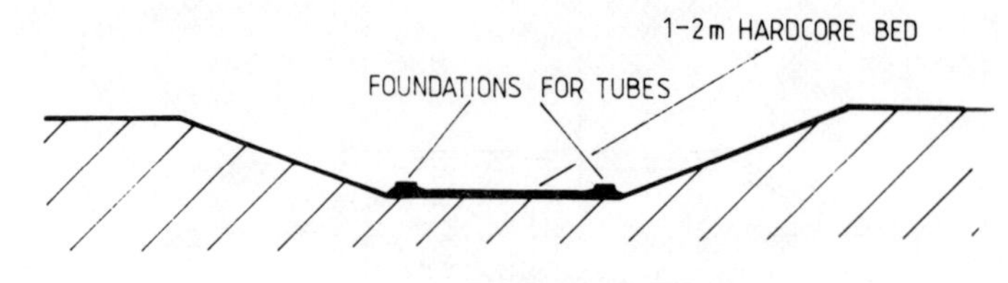

Fig. 10.54 Screeded bed for an immersed tube

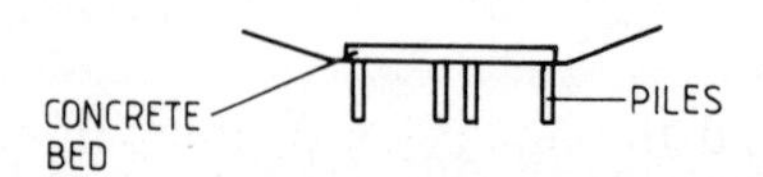

Fig. 10.55 Piled base for an immersed tube

(iii) sinking the tunnel units
(iv) sealing the joints

Reinforced concrete units are partially constructed on dry land, temporarily sealed at each end, launched into the water and built up to full dimension and towed into position. A unit is attached at each of the four corners to pontoon-mounted winches (Fig. 10.56). Ballasting is

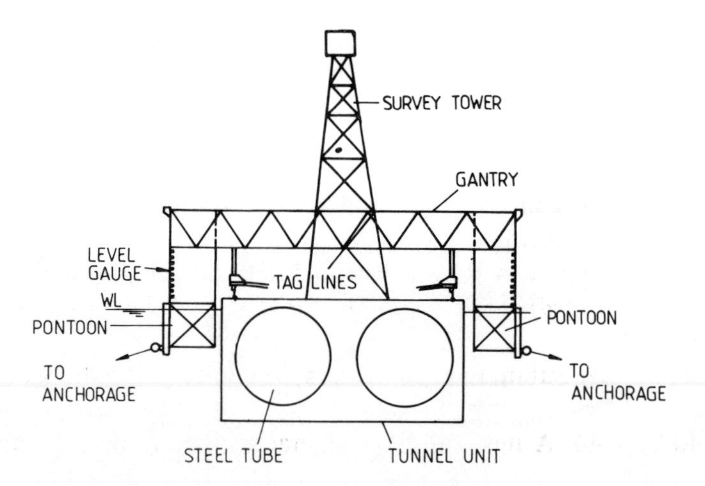

Fig. 10.56 Pontoon-mounted winches for tube positioning

subsequently increased to cause a loss of buoyancy and the section is lowered into place on the foundation pads under winched control. The final position is checked by divers and adjusted if necessary. The joint between the segments is subsequently sealed with concrete, placed with the aid of divers. To reduce the risk of movement, the space between

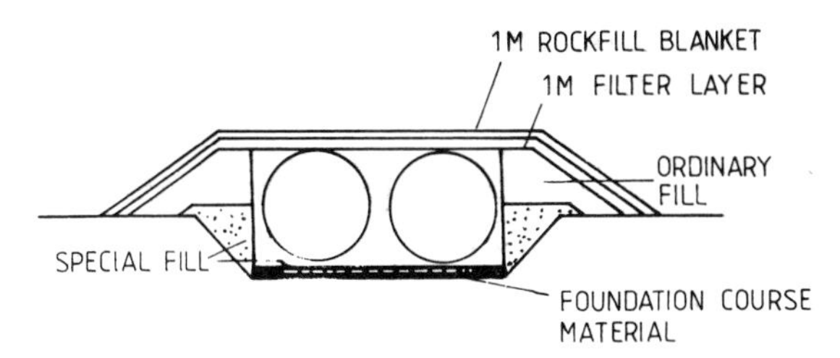

Fig. 10.57 Protective covering for an immersed tube

the hard core bed and the underside of the unit should be filled with coarse sand (Fig. 10.57), pumped from barges.

An alternative method may be chosen, using steel tubing supported on a reinforced concrete keel and surrounded by a thin concrete shell to provide protection. The unit is sunk to position as described above, and the joint temporarily sealed with concrete on the outside. The space between the two diaphragms of the new and the previously placed unit is pumped out and the end diaphragms are removed, leaving only the seal at the other end of the newly placed tube. The units can then be welded together in the dry.

A final precaution for either method, particularly near heavy shipping, is the placing of a protective covering (see Fig. 10.57).

Bibliography

Aleman, V.P. (1981) 'A strata index for boom-type road headers'. *Tunnels and Tunnelling*, **13** (2), March.

Bartlett, J.V., Biggert, A.R. & Triggs, R.L. (1973) 'Bentonite Tunnelling Machine'. *Proceedings of the Institution of Civil Engineers*, London, Paper **7577**.

Brakel, J. (1978) Submerged tunnelling, Technische Hogeschool, Delft, Holland.

CIRIA Report 44. A medical code of practice for work in compressed air.

Clarkson, T.E. and Ropkins, J.W.T. (1977) 'Pipe jacking applied to large structures'. *Proceedings Part I, Institution of Civil Engineers*, London, Nov.

Courtburn Ltd., Grundamet subsoil displacement mole. Oakley, Bedford.

Deere, D.U. et al. (1969) Design of tunnel liners and support systems. US Dept. of Transportation, (NTIS PB 183799).

Gosney, J. (1976) 'Bentonite tunnelling – the soft ground option'. (Summary of papers.) *Contract Journal*, **270**, 11 March.
Hammond, R. (1959) *Tunnel Engineering*. Heywood & Co.
Hartman, H.L. (1961) *Mine Ventilation and Air Conditioning*. The Ronald Press Co., New York.
Holdo, J. (1980) 'Tunnel driving techniques until the year 2000'. *Consulting Engineer (GB)*, **44** 5, May.
Hough, C.M. (1978) 'Pipe jacking case histories'. *Tunnels and Tunnelling*, **10**, April.
Jeffery, A.H.G. (1962) *Immersed tube tunnels*, Kemp International Publications, London.
Leins, W. (1972) *Tunnelbau*. Aachen Technische Hochschule, Germany.
MacFeat-Smith, I. (1978) 'Quantification of the cutting abilities of road header tunnelling machines'. *Tunnels and Tunnelling*, **10**, Jan/Feb.
McGaw, T.M. & Bartlett, J.V. (1982) *Tunnels – Planning, Design and Construction*. Vols. **I** & **II**. Ellis Horwood Ltd., Chichester. (This book contains a comprehensive bibliography.)
Medical Code of Practice for Work in Compressed-Air. Medical Research Council Decompression Sickness Panel 1975 (often called the Blackpool Decompression Tables, prepared by H.V. Templeman of the Royal Naval Physical Laboratory).
Morgan, F.F.L. (1959) The growing importance of roof bolting. Machy Lloyd, **31**.
Pequignot, C.A. (1963) *Tunnels and Tunnelling*. Hutchinson Scientific and Technical, London.
Proctor, R.V. & White, T.L. (1977) *Earth Tunnelling with Steel Supports*. Commercial Shearing & Dumping Co., Ohio, USA.
Richardson, M. & Scruby, J. (1981) 'Earthworms' system will threaten conventional tunnel jacking'. *Tunnels and Tunnelling*, **13** 3, April.
Robbins, R.J. (1976) 'Mechanised tunnelling – progress and expectations'. *Tunnels and Tunnelling*, **8**, May.
Roxborough, F.F. & Rispin, A. (1973) 'The mechanical cutting characteristics of lower chalk'. *Tunnels and Tunnelling*, **5** 1, Jan – Feb.
Schach, R. et al. (1979) *Rock Bolting: A Practical Handbook*. Pergamon.
Stack, B. (1981) *Handbook on Mining and Tunnelling Equipment*. Wiley.
Steenson, H.N. (1974) 'Fast set shotcrete in concrete construction'. *Journal of the American Concrete Institute*, **71**, June.
Szechy, K. (1973) *The Art of Tunnelling*. Akad Kiado (Hungary), 2nd rev. edn.
The Work in Compressed Air Special Regulations 1958. HMSO, London.

The Work in Compressed Air (Amendment) Regulations 1960, HMSO, London.

Tube Headings Ltd. Internal company literature on pipe jacking, thrust boring and mini tunnelling.

Wahlstrom, E.E. (1973) *Tunnelling in Rock.* Elsevier.

Index